Networking Humans, Robots and Environments

Edited By

Nak Y. Chong

Japan Advanced Institute of Science and Technology
Japan

CONTENTS

FOREWORD

When I began my research in multi-robot systems over two decades ago, I had many discussions with prominent roboticists who thought that it was preposterous for anyone to attempt work on multi-robot teams. "Why explore *multiple* robot teams when we don't even yet understand how to build a *single* intelligent robot?", some said. Others said: "We will *never* use more than one autonomous robot for challenging applications such as space exploration – it is just too hard."

Clearly, time has proven these naysayers wrong. Robot team capabilities have now evolved from swarms of homogeneous agents that exhibit emergent group behavior to heterogeneous teams that intelligently exploit the unique competences of each team member. The potential of these robot teams has been successfully demonstrated in many laboratory and real-world applications around the world. In fact, the advances in multi-robot cooperation have been so dramatic in the past decade that many aspects of multi-robot control are now nearly solved problems.

Of course, the story does not end there. Progress in the domain of multi-robot teams has led to a new frontier – a world in which robots no longer work only with other robot teammates that are in close proximity. Instead, these latest robots are now beginning to interact with each other, with humans, and with their environments, using the power of the *network*. These networked systems of interacting entities can take advantage of information not only in close physical proximity, but also information that is within networked communication proximity. This opens up a broad spectrum of new possibilities for networked humans, robots, and environments. In effect, the network extends the knowledge and reach of the individuals in the system, enabling them to have a more wide-reaching impact on the space in which they operate, and to have increased situational awareness of their surroundings and of the entities that share that space.

The research community has recognized the importance and potential of networked humans, robots, and environments, evidenced by the numerous research projects, workshops, symposia, technical committees, and special issues devoted to this topic over the past decade. This is a quickly growing community, which is achieving both theoretical and practical advances in the field. We now have examples of networked entities working together in outdoor urban environments, in home and office spaces, and in medical applications, just to name a few. Clearly, progress is being made.

Achieving this progress, however, requires answers to many new research questions regarding the networking of humans, robots, and environments, several of which are addressed in this book. First, what are the opportunities presented by networked robot systems (Ch. 1)? How do we model and design spaces that include networked humans, robots, and environments (Ch. 3, 7, 8, 9)? How should the networked sensors and robots interact (Ch. 2, 4, 5)? How should the network itself be designed to facilitate this interaction (Ch. 6)? How do humans interface with networked robots (Ch. 10, 11)?

This book presents interesting ideas for addressing these questions, and sheds light on new opportunities in this realm. While much remains to be studied, this book makes important advances that help us to better understand how to achieve highly productive, networked interactions between humans, robots, and environments.

Lynne E. Parker

Computer Science University of Tennessee
Knoxville
USA

PREFACE

How do robots interact with each other and their environment? How do they interact with humans? Questions abound when robots form a constituent of our daily life and natural environment. This book dives into the heart of how to design a novel distributed robot control and interaction architecture for achieving autonomous decision making and task execution in dynamic, human-in-the-loop environments. Over the past decades, a variety of architectures have been proposed for generating and controlling intelligent behavior of robots within a single robot platform mostly under centralized control. Responding to the growing need for large-scale service robotics applications under varying environmental conditions, new paradigms for the coordination and control of robots have been required in recent years. This book will examine several possible paradigms and explore their distinguishing features that enable robots to create emergent cooperative behavior in a fully or human-assisted semi-autonomous way. Those paradigms are investigated and analyzed on the theoretical grounds and experimental confirmation under real world conditions. As standalone autonomy is still more of a far-reaching goal, we shift our focus of attention toward networked autonomy. Provided that the compatibility and interoperability of network elements are guaranteed, networked autonomy could be considered to be more efficient and scalable at implementing various sophisticated robot applications than standalone autonomy. It will also exhibit high potential for future scientific breakthrough related to inter-robot and human-robot interaction and cooperation.

This book focuses primarily on latest techniques that proved successful for the control of robots in the current and near future networked environments, as well as try to bring to light difficulties that traditional standalone control approaches have faced due to lack of information available through the network. It has recently been demonstrated that a pervasive network environment can be used to give robots the ability to integrate information from diverse data sources both within and outside their own platforms into a single coherent framework. An integrated information infrastructure can then be built and shared by robots cooperating with each other and/or with humans. Specifically, the contributors discuss the principles and techniques that enable distributed, yet structured information to control the behavior of robots in the real world in more efficient and systematic ways. A notable example of network-based smart infrastructures includes cyber-physical systems that integrate the system's computational and physical resources. Along these lines, this book will show some of the most recent and breathtaking examples of inter-robot and human-robot networks operating in real environments. I believe that those examples will help readers understand and examine how and why various forms of cooperative interactions emerge and flourish in the robot-robot and robot-human environment.

Specifically, when robots are expected to exhibit some form of intelligence, they must be provided with information with regard to the state of the world, either at a local or global level, from which they make the best possible decisions about executions. The traditional approach integrates *a priori* information presumed by users or programmers, as well as acquired information through onboard sensing, to aid the robot's decision making process. It overburdens processing power, as entities in the world increase in complexity and require a vast amount of storage of data. Ubiquitous computing technologies provide a more convenient and efficient environment that facilitates the reshaping of the current monolithic architecture of behavior control. If the world becomes structurized in the sense of information management by tagging with appropriate information, robots will be effortlessly connected to the intrinsic features of the world. Moreover, if robots are aware of all the changing states of the world, then they will be able to exhibit an autonomous intelligence. The technical challenge is 1) how to build a network infrastructure that will let robots share information about the world in which they exist, 2) how to develop flexible control architectures and networking technologies for coordinating the actions of large numbers of distributed sensors and devices interacting with the world, 3) how to attain control over all the states of the entities in the world, and 4) how to automate the process of controlling behavior of robots in the world pre-populated with various types of information. The features of the coordinated control and information architectures proposed in this book will push robots toward achieving greater autonomy and decision that enable them to configure and adapt their behavior on demand in accordance with the directions of information management embedded in the surrounding environment. Those proposed architectures can be readily adopted in dynamically varying environments, helping ease the cost and complexity of real-time planning and execution of robot missions. Furthermore, robots can respond more effectively to hidden or explicit intentions of human interaction. Practical strategies for implementation build upon such emerging areas as ubiquitous computing, ambient intelligence, smart environments, distributed systems, wireless sensor networks, ontology, cloud computing, and so on.

This book will be of interest to the academic and industry professionals from the field of network control architecture, distributed multi-agent systems, artificial intelligence, behavior generation and control, robot-robot and human-robot interactions, and network applications design, whose attention is devoted specifically to the integration of heterogeneous information systems.

However, the author confirms that this eBook has no acknowledgment and conflict of interest.

Nak Young Chong
Japan Advanced Institute of Science and Technology
Japan

List of Contributors

Gerard McKee

School of Systems Engineering, University of Reading, Whiteknights Campus, Reading, Berkshire, RG6 6AY, United Kingdom

Gerard McKee

Faculty of Computer Science and IT, University of Baze, Abuja, Nigeria

Dezhen Song

Department of Computer Science and Engineering, Texas A&M University, College Station, TX 77843, USA

Fulvio Mastrogiovanni

Department of Computer, Communication and System Sciences, Faculty of Engineering, University of Genova, Via Opera Pia 13, Genova 16145, Italy

Geunho Lee

School of Information Science, Japan Advanced Institute of Science and Technology, Ishikawa, Japan

Kevin Lynch

Department of Mechanical Engineering, Northwestern University, Evanston, IL 60208, USA

Lichuan Liu

Department of Electrical Engineering, Northern Illinois University, DeKlab, IL 60115, USA

Takeshi Sasaki

Institute of Industrial Science, The University of Tokyo, Tokyo 108-8548, Japan

Bong K. Kim

Intelligent Systems Research Institute, National Institute of Advanced Industrial Science and Technology (AIST), Tsukuba, Ibaraki 305-8561, Japan

Wonpil Yu

Robot Research Department, Electronics and Telecommunications Research Institute, Daejeon, Korea

Dongjun Lee

Department of Mechanical, Aerospace & Biomedical Engineering, University of Tennessee, Knoxville, TN 37996, USA

Jee-Hwan Ryu

School of Mechanical Engineering, Korea University of Technology and Education, Cheonan, Chungnam, Korea

Nak Y. Chong

School of Information Science, Japan Advanced Institute of Science and Technology, 1-1 Asahidai, Nomi, Ishikawa 923-1292, Japan

KEYWORDS

Chapter 1: Networked Robotics: Developments and Opportunities

Gerard McKee and Blesson Varghese

Keywords: World Wide Web, wireless network, networked robotic system, online robot system, distributed communication model, multi-level distributed robot architecture, swarm-array computing, teleoperated system, distributed robotic system, hierarchical framework, loosely coupled coordination, tightly coupled coordination, explicit communication, implicit communication, multi-team multi-robot system, collision avoidance, pattern transformation, peer-to-peer communication, parallel reduction algorithm, fault tolerance

Chapter 2: Collaborative Crowd Surveillance Using Networked Robotic Cameras

Yiliang Xu and Dezhen Song

Keywords: networked pan-tilt-zoom camera, fixed wide-angle camera, collaborative observation, camera coverage, camera zoom level selection, camera traveling time, frame selection problem, master-slave camera configuration, camera scheduling, network packet scheduling, scalability, non-parametric Gaussian background subtraction, Kalman filter, least overlapping condition, non-overlapping condition, coverage-resolution ratio, lattice-based induction-like algorithm, approximation bound, branch and bound-like approach, lattice pruning scheme

Chapter 3: Distributed Sensing and Human-Aware Robot Reasoning Mechanisms

Fulvio Mastrogiovanni, Antonio Sgorbissa and Renato Zaccaria

Keywords: Formal language, ubiquitous robotics, distributed sensing, skilled actuation, situatedness, human-robot interaction, distributed context assessment, knowledge representation, human activity recognition, symbolic structure, semantic relationship, situation definition language, context toolkit, context broker architecture, aspect-scale-context, context ontology language, cooperative anchoring, associated inference, context-aware behavior, artificial cognitive system

Chapter 4: Self-Configurable Mobile Robot Swarms: Adaptive Triangular Mesh Generation

Geunho Lee and Nak Young Chong

Keywords: sensor network, swarm robot, environmental monitoring, area coverage, self-configuration, local interaction, scalability, collective behavior, triangular lattice, partially-connected topology, sensing boundary, neighbor robot, Lyapunov theory, centroid, orthocenter, equilibrium state, stability, spatial density

Chapter 5: Experimental Validation of Multi-Agent Coordination by Decentralized Estimation and Control

Michael Hwang, Matthew L. Elwin, Peng Yang, Randy A. Freeman, and Kevin M. Lynch

Keywords: Decentralized framework, formation control, information diffusion, inertia moment, local sensing, local communication, group behavior, dynamic average consensus estimator, motion controller, discrete-time estimator, uniform density, gradient control law, collision avoidance filter, saturation stage, indoor positioning system, XBee radio, limited communication radius

Chapter 6: Extending Lifetime of the Network and Crucial Node by Multiple Diversity Combining

Lichuan Liu

Keywords: crucial node, sensor network, energy consumption, network lifetime, packet collision, collision separation, space and network diversity combining, media access control layer, cross-layer design, antenna array, multi-hop routing, slotted random access, network assisted diversity multiple access, linear inverse filtering, goodput analysis, delay analysis

Chapter 7: Intelligent Space: A Platform for Integration of Robot Technology

T. Sasaki and H. Hashimoto

Keywords: intelligent space, human intention, distributed intelligent network device, sensor node layer, RT middleware, OpenRTM-aist, object tracking,

Kalman filter, sensor fusion, automated sensor calibration, spatial memory, spatial knowledge tag, human activity recognition, hierarchical clustering, mobile robot navigation, Field D* method, dynamic window approach, occlusion avoidance

Chapter 8: Universal Design of Ubiquitous Robotic Space

Bong Keun Kim, Hyun Min Do, Hideyiki Tanaka, Yasushi Sumi, Hiromu Onda, Tamio Tanikawa, Kohtaro Ohba and Tetsuo Tomizawa

Keywords: ubiquitous robotic space, universal design, ubiquitous function service, service-oriented architecture, smart object, smart logic, smart discovery, ubiquitous function activation module, robust internal-loop compensator, reproducible structure, u-RT space, RFID, reconfigurable control method, librarian robot, daily-life-supporting robot, coded landmark for ubiquitous environment, asynchronous java script and xml technology

Chapter 9: Ubiquitous Robotic Space and Its Real-world Applications

Wonpil Yu, Jae-Yeong Lee, Heesung Chae, Yu-Cheol Lee, Minsu Jang, and Joo-Chan Sohn, Hyosung Ahn Young-Guk Ha and Yong-Moo Kwon

Keywords: ubiquitous robotic space, robotic service framework, ambient intelligence, distributed sensing, situation understanding, service generation, physical space, semantic space, virtual space, data communication interface, localization network, optical tracking scheme, starLITE, uClips, context aware service execution, security robot, mobile robot navigation, standardized service platform

Chapter 10: Toward High-Performance Stable Haptic Teleoperation over the Internet: Passive Set-Position Modulation Approach

Dongjun Lee and Ke Huang

Keywords: Passive set-position modulation, haptic teleoperation, varying communication delay, packet loss, stability, passivity, human dynamics, scattering/wave-variable formulation, Parseval's identity, PD control, virtual coupling, Colgate's passivity condition, passivity observer, passivity controller, Lagrangian dynamics, bounded realness

Chapter 11: Stable Teleoperation with Time Domain Passivity Control

Jee-Hwan Ryu, Dong-Soo Kwon and Blake Hannaford

Keywords: bilateral teleoperation, transparency, stability, time-domain passivity, energy-based method, passivity observer, passivity controller, haptic interaction, multi-port network, admittance causality, impedance causality, energy dissipation, damping distribution, five-bar mechanism, bilateral position-force control, PD position control

2

CHAPTER 1

Networked Robotics: Developments and Opportunities

Gerard McKee

Faculty of Computer Science and IT

Baze University

Abuja, Nigeria

and

Blesson Varghese

Faculty of Computer Science

Dalhousie University

Halifax, Nova Scotia, Canada

ABSTRACT. Networked Robotics has come a significant way since its inception in the early 90's in the form of online (web-based) robot systems. Its progress since that time has been allied closely to the development of the web, the Internet and computer networking technology, notably wireless networks. Where is networked robotics now and what does the future hold? This chapter explores these questions from the viewpoint of research in the Active Robotics Laboratory at the University of Reading in the UK for the period from the early 1990s up to 2011. This research has focused on networked robotics as a distributed set of computing and robotics resources. Previous work on online robot systems, distributed software models and cooperative robotics is reviewed along with recent work on swarm systems and swarm-array computing, which takes some of the ideas of networked robotics into the field of fault tolerance for high performance computing systems.

1. Introduction

Networked robotics combines the areas of robotics and computer networking to provide new opportunities for a range of network-based robotic systems including online robots and distributed and cooperating robots. In the first case, online robots, robotics combines with the Internet and World Wide Web technology to provide remote access to command and control robots at a distance. In the second and third cases, distributed and cooperating robots, robotics takes advantage of wireless communication networks to provide a platform for distributed communicating cooperating robots.

At the University of Reading we explored the integration of computer networking with robotic systems to develop networked robotic systems. We also more recently took some of the concepts of networked robotics into the area of high performance computing by way of swarm robotic systems. In this chapter we review some of this work and discuss these recent developments in detail.

The chapter is organised as follows. The following section provides a brief review of work in the area of online robot systems. We propose that there is still much work to be done to bring online robot systems into teaching programs and provide sustainable online robot laboratories. Following this are two sections on multi-robot systems. The first discusses a distributed communication model for effectively wiring together geographically distributed robotics modules to form multi-level distributed robot architectures. The second discusses tightly coupled cooperating robot systems where communication between the robots is both implicit and explicit. A further section introduces our recent work on swarm robotic systems, specifically addressing communication requirements for command and control of the robot swarm. This work led to a number of ideas for the exploitation of distributed communicating agents to support fault tolerance in high performance computing systems, resulting in the concept of swarm-array computing. The penultimate section of the chapter discusses swarm-array computing. The final section of the chapter provides a summary and conclusions.

2. Online-Robot Systems

The Mercury project was one of the first online robot systems closely followed by the Bradford Telescope project [1] [2]. Both exploited the newly emerging World Wide Web to provide in the first case web-based control of a robot arm and in the second case the ability to point and capture images of the night sky with a telescope. The idea of an online robot is apparent from these two examples, for it enables users located remote from the robot to both issue commands to the robot and to get feedback from the robot and its environment.

Online robot systems can be characterized as either open or closed, depending on whether they are open to general members of the public or reserved for use by students on taught courses. A good example of an open system is the robot TeleGarden, developed by Ken Goldberg and his students at University of Southern California [1]. Another more recent example is the Red Rover online demonstration championed by The Planetary Society to support outreach prior to and during the Mars Exploration Rover (MER) missions to Mars in 2004 [3].

A good example of a closed system was a simple robot arena located in the Active Robotics Laboratory at the University of Reading. The arena incorporated a toy digger which could be controlled from a server and onboard and arena-mounted cameras which were connected to frame grabbers on the same server [4]. Students could remotely login to the server to join a queue for control of the robot and grab images from the camera. The task was to first construct a graphical user interface to remotely teleoperate the digger to pick up an object in the arena and then to develop an artificial intelligence program integrating vision sensing with robot control to automate the pickup operation. The assignment was repeated each year with 80-100 students on each occasion.

A more recent example of a closed system which is well integrated into a University program is provided by Raul Marin's work at University of Jaume I, Castellon, Spain [5] and the Bradford Telescope continues to go from strength to strength [6]. A good example of a system which has spanned both open and closed domains is PumaPaint, which was initially set up as a means for computer science students in one institution to develop graphical user interfaces to control a Puma robot located in a remote laboratory, and was then opened out to public use [7].

Online robots of the type presented above are essentially teleoperated systems with the communication link between the user controls and the robots via a computer network rather than dedicated cabling. Since the internet introduces variable delays the above systems typically adopt a stop-look-act cycle, whereby the robot stops, sensor data is collected, an operator or an artificial intelligence program make a decision to act based on interpretation of the sensory data and then waits for the action to be completed before repeating the cycle.

Online robots have also found application in commercial settings. One prominent example is telesurgery, in which a surgeon in one geographical location may remotely operate a robot via a computer network to carry out an operation on a patient in another geographical location [8]. The Zeus™ and the daVinci™ robot systems have been demonstrated in this online robot mode [9] [10]. For example, the Zeus system was used in a transatlantic telesurgery demonstration over an ATM network to ensure low latency [9].

One area of confusion that has emerged with the creation of online talking heads is the meaning of an online robot for these talking heads are sometimes referred to as online robots.

One possible area of future development which hasn't yet been explored is the creation of educational programs in robotics centred on the use of online robot systems, enabling students and general members of the public to study robotics. This requires thinking about robotics education programs outside of the normal degree program setting. These educational modules could be integrated with off-the-shelf robot toolkits such as the Lego®Mindstorms®robot kit [11] to provide an extended program of robotics education. In this context, the robot arenas must be designed for online use, which implies in the first instance that the arena is permanently live and self-maintaining and in the second instance that the arena provides the functionality required for the educational outcomes that are to be achieved. The latter in particular requires careful design of the robot system and its environment to avoid damage and maintain consistent operation.

The commercial exploitation of online robot systems has yet to be fully developed. While telesurgery has offered demonstrations of online robot based surgery, the communication requirements for low latency delay poses challenges to its widespread use. One area of potential use in a commercial setting is in technology outreach, and one notable application area is the space industry. In 2004, NASA carried out field trials with Mars exploration rover prototypes to enable scientists to work with the robot technology prior to the mission to Mars [12]. When the scientists had completed their work school children were brought into the command centre and allowed to command-and-control the robot, which was located in the Mojave desert. The Red Rover project, although based on toy robots also offered a level of outreach to engage students and members of the public in scientific exploration [3]. A further example is the JASON project, which offered school children the ability to interact with a remote submersible robot [13] [14].

3. Distributed Robotic Systems

The term 'distributed robotic systems' traditionally refers to a set of geographically distributed mobile robot systems which may exploit wireless communication to share information and coordinate tasks. We adopt a more inclusive definition in which the robotic systems may comprise traditional mobile robot systems in the large, for example mobile robot systems for outdoor surveillance, in the small, for example swarm robotic systems or distributed mobile sensing units, and fixed robotic modules including geographically distributed sensors and actuators as in wireless sensor and actuator networks. We adopt in addition a modular service-oriented viewpoint in which the sensors and actuators within a traditional robot system, and these fixed sensor and actuator systems are modules offering a range of functionality and connectivity, and possess both geographic and network location.

The approach we adopt can be likened to the multi-agent systems approach for it requires software components that can listen and respond to queries [15], offer services to other modules and in the ideal case some will also possess mobility so that they can relocate to network nodes with free resources or where they are closer to modules with which they are directly connected. However, multi-agent systems research tends to focus on agents that are autonomous with proactive, reactive and social decision making and interaction. The approach we have followed treats the modules more as the components of a system, with each module providing services to one or more other modules, and all modules working in their own way towards the operation of a distributed robotic system. In this respect it is more appropriate to think of the modules as being wired together into a multi-level architecture even though the modules themselves are geographically distributed.

We explored this approach with the development of DINERO (Distributed NEtworked RObotics) [16]. Figure 1 shows a diagrammatic depiction of the DINERO framework. DINERO assumes the presence of a pool of modules that are geographically distributed across multiple mobile robot platforms. Modules are self-aware, incorporating descriptions of their location and capabilities, and network-aware, able to accept and initiate network-based communication with other modules. Modules are either resource modules or task modules. Resource modules provide services. Task modules integrate resource modules to create higher-level services. Task modules are essentially un-instantiated resources which glue together resource modules to create these higher level services. In order to become instantiated as resources, task modules exploit a query-based framework

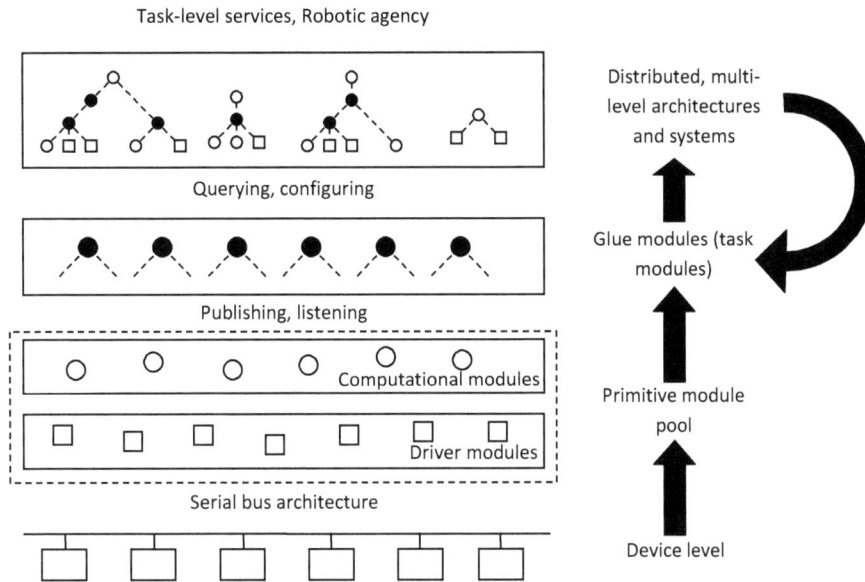

FIGURE 1. The DINERO framework provides a hierarchical framework for integrating resources to create higher level resources and networked robotic systems

to locate resource modules that satisfy specific functional and non-functional requirements. The non-functional requirements specify, for example, the geographic location of resource modules and their relationship to other modules. Resource modules listen to and respond if they satisfy the requirements embodied in the queries. When a task module has located the resources needed to enable it to offer itself as a higher level service it links with the resources, effectively wiring them together, and then either pursues a task as a networked robotic agent or listens for requests for its services as a resource module. Robot architectures, therefore, are built in a hierarchical manner from the bottom-upwards, incorporating successively higher-level task modules.

The DINERO framework offered initial insights into the structuring of software to support distributed cooperating resources. Further work is now required to redesign the DINERO framework taking into account the existing experience with DINERO, particularly the software architecture of the software modules, the language for describing constraints, and a wider range of case studies. In addition, further work is required on the provision and distribution of computational modules, including a module server that can inject modules onto the network to support the requirements of task modules; and with this the ability for computational modules to be mobile. Further work is required in support of explicit communication between robotic agents, possibly exploiting communication languages developed by the multi-agent systems research community. Finally, a further area of exploration is the integration of methods and approaches across online robot systems, distributed and cooperating robots, and intelligent spaces (in this book), exploring further the integration of fixed and mobile robotic agents and components in wireless sensor and cognitive sensor networks, and indeed the interfaces between these and humans.

4. Cooperating Robot Systems

The term 'multi-robot system' can be used to refer to a wide range of robotics systems incorporating more than one robot, including swarms of many robot systems and smaller robot teams for competitions such as Robocup [17]. The term is used in this section to refer to heterogeneous teams of mini or larger mobile robot systems, including the teams for example in the Middle Sized league of RoboCup. Such systems typically incorporate wireless Ethernet as the basis for communication, vision based sensing, an on-board computer, possibly a laptop, and hence the ability to support a greater level of autonomy as well as robot-robot and human-robot cooperation.

There are four general challenges that can be identified in the research literature. First, how do we ensure that the robots stay within range of each other so that a communications network is maintained encompassing all the robots (e.g. [18] [19]). Second, in a cooperative sensing or surveillance task, how do we ensure that the area that is to be surveyed is adequately covered (e.g. [20]). This includes elements of target allocation, in that each robot needs to be allocated or determine an area to be monitored. Third, in a cooperative mapping task, how is the information from the local maps built by each robot integrated to provide a global map [17]? Fourth, in a robot team that is to perform some task that can be broken out into subtasks, how should the sub-tasks be allocated to the robots in the team (e.g. [21] [22])?

Cooperating multi-robot systems can generally be classified as loosely coupled multi-robot systems since the robots are associated in the completion of a task but are not physically coupled to other members of the robot team. In contrast, robot systems which are involved for example in the cooperative transport of extended payloads and are therefore physically coupled by means of the object being transported can be classified as tightly coupled multi-robot systems. In the first case the communication between the robots is largely explicit, network-based, while in the second case the robots can take advantage of implicit communication via the connecting payload.

Tightly coupled operation of multiple robot systems is required when two or more robots are working together to manipulate an extended object. These studies cover work in the area of dual manipulator systems (e.g. [23] [24]) and in mobile robot mounted manipulator systems [25] [26] [27] [28]. In this section we will focus on cooperating multi-robot systems for carrying an extended payload, reporting in particular the research carried out at the NASA/CALTECH Jet Propulsion Laboratories and in the Active Robotics Laboratory at the University of Reading.

Research in the area of multiple robots collectively carrying a shared object has been pursued in the context of space robotics missions, specifically scenarios in which a Habitat is prepared for human arrival on Mars [26]. A particular task associated with the preparation of a human habitat on Mars involves the deployment of PVC tents to harness power from the sun. The scenario involves a container storage unit (CSU) containing a set of PVC tents and a cooperating pair of robots to (a) approach and pick up individual PVC units from the CSU, (b) transport the units to the target site, and (c) set down the units at the destination site [29]. This is repeated until all units are deployed.

The research in the Active Robotics Laboratory at the University of Reading, illustrated in figure 2 [27] focused on the overall design of the dual robot system to support the task, including pickup, traverse and set down, and takes its inspiration from work carried out at the UCLA/NASA Jet Propulsion Laboratory [26]. Both sets of work assume a hybrid architecture, in JPL's work this is referred to as the CAMPOUT architecture [29], combining deliberate top-down planning of the task and lower level behaviour-based control for

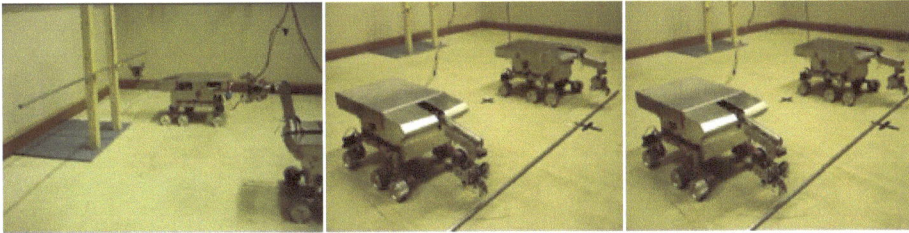

FIGURE 2. Three stages in a cooperative transport task - approach to pickup (left), transport (centre), and putdown (right).

immediate robot-robot interaction during transport. However, neither has implemented the deliberative component, opting to focus on designing the behaviour-based coordination and defining the overall task structure manually. Both have employed implicit communication based on sensing the motion of the payload in the grippers when cooperatively carrying the payload, though the JPL work has implemented also explicit communication between the robots for synchronisation. Both sets of work demonstrated practical systems and the JPL work was successfully demonstrated in outdoor terrain. The work at the University of Reading focused on specific behavioural aspects of the architecture, opting to manually set and unset behaviour sets for each component of the task [**27**].

An area of future work is to explore multi-team multi-robot systems. A flexible and robust multi-robot system must incorporate within individual robots and across the team the heterogeneity needed to support a wide-range of tasks. This is an important challenge for multi-robot systems research and therefore has implications for networked robotics. This will require providing support for explicit communication and coordination between the robots in a team. In addition a number of robot teams may be formed within a multi-robot system, and even within one or more of these teams further structuring as sub-teams may be required. It should also be possible for robots to move between teams as the requirements on a set of tasks changes. Explicit communication is then required not only between individual robots but also between robot teams.

A space robotics application involving the cooperative transport of an extended payload can be used to illustrate the multi-team multi-robot concept. Assume that a heterogeneous pool of robots is provided from which robot teams can be drawn. The selection process involves the identification of roles and then the allocation of robots to the teams that fulfil these roles. In the transportation task robot teams are required to respectively carry an extended payload, clear a path across the terrain for the transport of the payload and scout a route to the destination. These three teams can be called the 'carriers', the 'clearers' and the 'scouters' respectively.

The scouters will need to coordinate with each other to explore and select a suitable path through the terrain between the origin and the destination. They will then need to communicate the selection to the clearers and also to the carriers. The clearers need to coordinate with each other to clear rocks and flatten the terrain along the path, keeping ahead of the carriers. Clearing the path may involve the formation of loosely and tightly coupled sub-teams within the clearing team. Tightly coupled coordination may be required for example to push aside a large rock. The clearers need to coordinate with the scouters and the carriers to confirm that the path (ahead) of the carriers has been cleared. The carriers are responsible for picking up the object to be transported, carrying it along a path and setting it down or assembling it at a destination. This is a tightly coupled operation requiring coordination between the robots in the team to allocate grasp points, synchronise on pickup and cooperate on traversal.

Each robot will need to ensure that it has the resources to perform its role and to evaluate satisfactory completion of its role. The robots within each team must also synchronise to ensure that the task assigned to its team has been satisfactorily completed. Moreover, there is a dependency between the teams since the scouters must provide the route that the clearers need to clear in preparation for the carriers to move the payload. However, these three activities can proceed in parallel provided the scouters task stays ahead of the clearers task and the clearers task in turn stays ahead of the carriers task. Within this setting it is also possible to envisage robots moving between teams as the level of difficulty of the three tasks changes, facilitating a still more flexible and robust multi-robot system.

5. Swarm Robotic System

One area where inter-robot communication is relevant is swarm robotic systems. Robots in these systems are minimalist and therefore simple communication techniques, short-range communication and economic messaging are the norm. However, within existing approaches to communication within swarm systems there is no systematic means to determine the communication requirements except at the level of individual messages. There is no means for example to determine how many messages may need to be communicated per second and the scope of individual communications - whether peer-to-peer, peer-to-group or group-to-group. This can be addressed if there was a mathematical model for swarm systems. However, a review of the literature revealed that such a model was missing. The lack of such a model motivated research reported elsewhere aimed specifically at the development of such a model [**30**] [**31**]. This model provides a framework within which to study the communication requirements of a swarm robotic system.

The mathematical model addresses a number of challenges [**32**] including obstacle avoidance, collision avoidance and transformation of swarm patterns. The model is based on the complex plane: swarm robotic agents are placed on the roots (vertices) of a complex equation to form a polygonal pattern. The parameters that define the mathematical model are classified as macroscopic and microscopic. Macroscopic parameters define group or abstract behaviours of the swarm system whereas microscopic parameters define individual agent behaviours. Many experimental test-beds for swarm robotic systems, for example Mission Lab [**33**] [**34**], are based on microscopic parameters. This means they focus on peer-to-peer communication between the robots.

Pattern transformation is a key aspect of swarm systems. Obstacle avoidance for example can be treated within swarm systems as pattern transformation. For example, a swarm of robots in a polygonal pattern needs to transform to a linear pattern to traverse through a narrow path and then transform back to the previous pattern on reaching the far side. The challenge of pattern transformation was addressed in the mathematical model with two feasible tools, the first a non-mathematical tool [**35**] and the second a mathematical tool [**36**]. The first was based entirely on macroscopic parameters while the second, which employed Moebius transformations, incorporated both macroscopic and microscopic parameters.

In both methods, attention has to be paid to ensuring that the robots do not collide with each other during transformation. The first method achieves this by first rotating the polygon by a predefined offset and then applying a macroscopic parameter adjustment to create a new pattern (figure 3, the left three images). If the new pattern is a line pattern the macroscopic parameter adjusted is the lateral radius of the circle in which the polygonal pattern is circumscribed. A further rotation is then performed to ensure a regular pattern; in

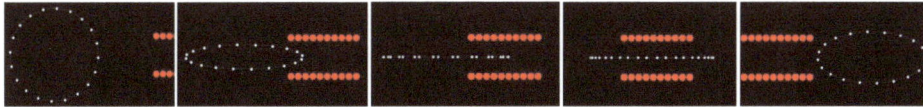

FIGURE 3. Circle-to-line and line-to-circle transformation of a robot swarm in order to navigate a narrow obstacle in the first method.

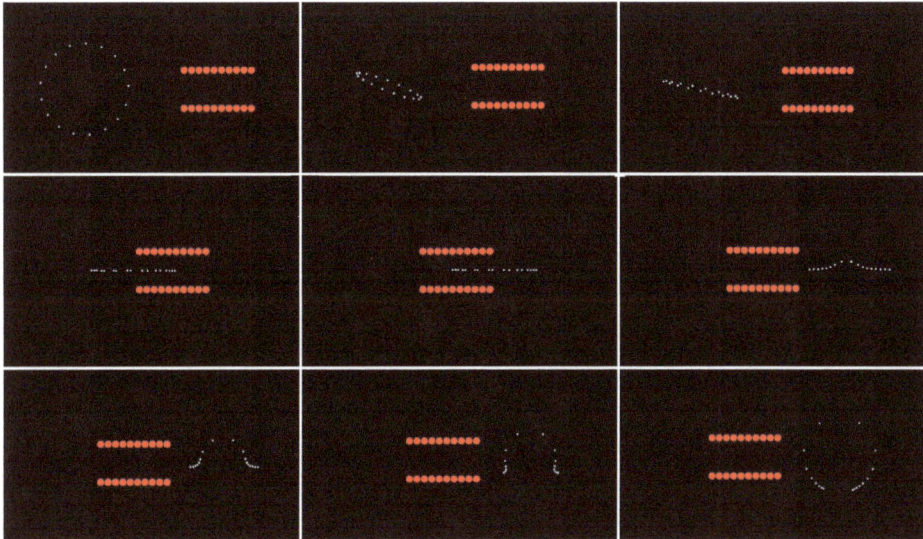

FIGURE 4. Circle-to-line and line-to-circle transformation of a robot swarm in order to navigate a narrow obstacle in the second method.

the case of a transformation to a line the further rotation effectively adjusts the separation between the robots to equidistant (figure 3, third and fourth images).

The second method avoids collisions between robots by first shifting from a global frame of reference to a local reference frame, performing a discrete (Moebius) transformation along with path planning to take the robots to their positions in the new pattern (figure 4, first row), and then finally shifting from the local to the global reference frame (figure 4, second row first image). Figure 4 shows the transformation from a circle to a line (first row and left image of second row) and then a line to a circle (right image of second row and third row). Microscopic parameter adjustments occur within the local frame of reference while macroscopic parameter adjustments occur within the global frame of reference. The microscopic parameter adjustments include the positions of the robots whereas the macroscopic parameter adjustments include for example the compactness of the pattern. The pattern resulting from the line to circle transformation in figure 4 is distorted since the robots are not distributed evenly around the circle; this is a limitation of the computer modelling of the transformation since only a discrete set of points (the robots positions) are used in the transformation.

The selection of the transformation tool impacts on the communication requirements for a swarm system. In the first method three broadcast messages are required. The first is to effect the first rotation, the second to effect the pattern transformation and the third to effect the additional rotation. These messages need only be a few bytes in length. In

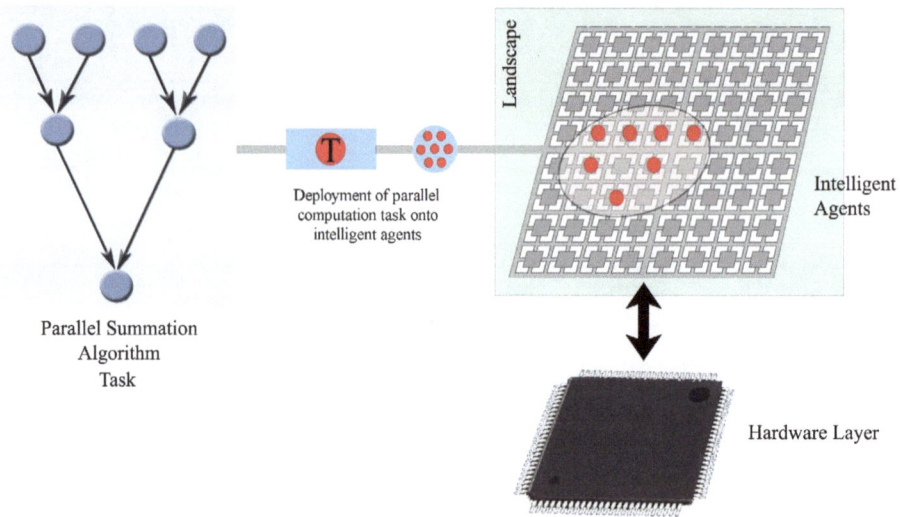

FIGURE 5. Swarm-Array Computing: the components of a parallel algorithm are carried onto computing space by swarm agents

addition, each robot will only receive three short messages, none of which it will need to discard. In contrast, in the second method, a centralised global planner is required which in practical terms means offloading the planning to a high-performance workstation. Separate plans need to be generated for each individual robot and then each plan needs to be communicated to its target robot. This means that each robot receives many communications only one of which it is required to accept and each communication is a large packet of information because it contains a discretized plan for the robot. In summary, therefore, a mathematical framework can provide a basis for determining the communication costs for swarm systems.

6. Swarm-Array Computing

Networked Robotics offers a unique viewpoint on some traditional problems in robotics, since it brings together networking and robotics technologies. It has also inspired a new way of looking at some problems outside robotics. For example, if a problem is solved using parallel algorithmic methods, the parallel components may be distributed across multiple computing nodes, and will require communication in order to coordinate if there are dependencies between the components. This model is very close to the distributed software model of networked robotics.

Parallel reduction algorithms are important in high performance computing [37]. They are based on tree structures and one particular form of algorithm involves data flowing from the leaves of the tree towards the root. At each intermediate node the converging data input is transformed into a result that is passed forward to the next intermediate node. Figure 5, left, shows an example. The interconnection of a node in the tree represents its dependencies. For a binary tree, which is a popular data flow structure in high performance computing, the root node and each intermediate node has two input dependencies. For an N-ary tree the dependencies increase with N. This means that the complexity of communication and coordination between the nodes also increases with N.

When a parallel reduction algorithm is run on a high performance computer each node of the tree is scheduled onto a separate computing node. An important issue in high performance computing is that these computing nodes are susceptible to failures. The isolation of these failures is dealt with under fault tolerant computing. The conventional approach to addressing failures is checkpointing, which involves recording intermediate states of execution to which execution can be returned if one or more nodes on which the algorithm is executing fails. The traditional method for detecting failure has been administrator-based, whereby the human administrator of the system detects that the algorithm has stalled.

Checkpointing in general has two major drawbacks. Firstly, if there is a single point of failure then either all the nodes are restarted from scratch (a cold restart) or from a previous checkpoint [37] [38]. Decentralized checkpointing allows finer control of the restart. Secondly, if there are large number of nodes and dependencies the checkpointing overheads will be large [37]. These all reduce the efficiency of high performance computing systems.

The efficiency could be improved if in some way the algorithm itself could be self-managing. A simple definition of self-management in this context would be where if a node is about to fail the component of the algorithm can be moved off the node and the input and output dependencies re-established on another node. This would require the individual components to incorporate some level of agent-like intelligence whereby the condition of the computing node can be monitored and the component moved if failure is predicted.

One approach to incorporating intelligence is to map all of the parallel components of the task onto a set of agents such that the algorithm is essentially the payload of the agents. Figure 5, middle, shows the agents. The set of agents then carry the payload onto the array of computing nodes. The array can be viewed as a landscape and the set of agents as essentially a robot swarm. The landscape comprises obstacles, which are nodes that have failed or are about to fail, and elevated terrain, where the height of the terrain reflects the load on the active nodes. The swarm can then move over this terrain to find an expanse where it can compute. Figure 5, right, shows a swarm that has been located on the computing array. Swarm pattern transformation methods [31] [32] can be employed both to avoid obstacles and navigate narrow terrain, but also to compact the swarm into a small area if needed.

This approach provides a distributed alternative to one of the important yet centralized tasks in high performance computing, namely the scheduling of algorithm components to nodes. Further, if one of the nodes on which the swarm is located fails, a local adjustment can be made by the swarm agent relocating to a nearby part of the landscape and re-instantiating its dependencies. These two benefits of the swarm approach to high performance computing, which we refer to as swarm-array computing, offers the potential to improve the fault tolerance, and therefore increase the efficiency of high performance computing systems.

These concepts have been investigated practically through both a simulation and implementation [40] [41]. The implementation employed a computer cluster with thirty four nodes. Message Passing Interface (MPI) [42] [43] implementations, namely Open Message passing interface (OpenMPI) [44] [45] and adaptive MPI (AMPI) [46], built on Charm++ [47], were used as the middleware for the implementations. A parallel summation algorithm with fifteen nodes was implemented using both the classical approach and the swarm-array computing approach. The main conclusion of the implementation studies was that the swarm-array computing approach improved fault tolerance as measured by the mean time taken for reinstatement of the algorithm if a node failed. A further conclusion was that the existing middleware for high performance computing does not easily support the swarm-array computing concept. Notably, a number of workarounds were required to realize the implementation.

7. Conclusion

Networked Robotics addresses a range of challenges at the interface between robotics and computer networking. These challenges focus on the delivery of a communications framework to support online, distributed and cooperating robot systems. Significant advances have been made in the exploitation of online robot systems for education and public outreach, but more needs to be done to create robust, usable systems that engage the interest of users. A significant body of work has been produced in the area of distributed and cooperating robot systems, but more needs to be done to demonstrate the benefits of networked robotics over and above the provision of a communications infrastructure. Networked robotics has a significant role to play in the creation of human robot intelligent spaces and ambient intelligence, combining embodied and embedded robotics technology with cognitive and wireless sensor networks. We are continuing to explore these challenges and have also taken some concepts of networked robotics into the area of fault tolerance for high performance computing systems.

Bibliography

[1] K. Goldberg, "The Robot in the Garden: Telerobotics and Telepistemology in the Age of the Internet," MIT Press, 2000.

[2] K. Goldberg and R. Siegwart, "Beyond Webcams: An Introduction to Online Robots," MIT Press, 2001.

[3] L. D.Friedman, L. A. Hyder, S. L. Klug, J. S. Oslick, G. E. Powell, E. L. Thomas, J. L. Vaughn, "Red Rover Goes To Mars: An Exploration Education Experiment for the Mars Surveyor," in the Proceedings of the 30th Annual Lunar and Planetary Science Conference, 1999.

[4] G. T. McKee, "An Online Robot System for Projects in Robot Intelligence," International Journal of Engineering Education, Vol. 19, No. 3, 2003, pp. 356–362.

[5] R. MarúŃ, P. J. Sanz, P. Nebot, R. Esteller, "The Internet-Based UJI Tele-Lab: System Architecture" in the IEEE International Conference on Systems, Man & Cybernetics, 2003, pp. 4904–4909.

[6] J. Baruch, "Bradford robotic telescope: education & public outreach: journey into space," Astronomy & Geophysics, vol. 48, no. 4, pp. 4.27–4.28, 2007.

[7] M. R. Stein, "The PumaPaint Project" Autonomous Robots, Vol. 15, No. 3, 2004, pp.255–265.

[8] R.M. Satava, "How the Future of Surgery is Changing: Robotics, Telesurgery, Surgical Simulators and other Advanced Technologies," Jurnalul de Chirurgie, Iasi, 2009, Vol. 5, Nr. 4, pp. 311–325.

[9] Jacques Marescaux, Joel Leroy, Michel Gagner, Francesco Rubino, Didier Mutter, Michel Vix, Steven E. Butner, Michelle K. Smith, "Transatlantic robot-assisted telesurgery," Nature, Vol. 413, September 2001, pp. 379–380.

[10] D. Stoianovici, R. Webster and L. Kavoussi, "Robotic Tools for Minimally Invasive Urologic Surgery," Chapter in 'Complications of Urologic Laparoscopic Surgery: Recognition, Management and Prevention', Dec 2002.

[11] Lego Mindstorms website: http://mindstorms.lego.com

[12] M. Hardin, "New Mars rover FIDO works like a dog," Universe, Jet Propulsion Laboratory, Pasadena, California, Vol. 29, No. 10, 1999.

[13] http://www.jason.org

[14] R. D. Ballard, "The JASON Project: Hi-Tech Exploration Promotes Students' Interest in Science," Technological Horizons in Education, Vol. 20, 1992, pp. 70–74.

[15] M. Wooldridge, "An Introduction to MultiAgent Systems, Second Edition" John Wiley & Sons, 2009.

[16] Duncan Baker, "Distributed Network Robotics", PhD Thesis submitted to the University of Reading, UK, 2006.

[17] T. Balch and L. E. Parker, "Robot Teams: From Diversity to Polymorphism", A. K. Peters, Ltd., 2002.

[18] S. Poduri, and G. Sukhatme, "Achieving Connectivity through Coalescence in Mobile Robot Networks," in the Proceedings of the 1st International Conference on Robot Communication and Coordination, Athens, Greece, 2007.

[19] T. Facchinetti, G. Franchino and G. Buttazzo, "A Distributed Coordination Protocol for the Connectivity Maintenance in a Network of Mobile Units" in the Proceedings of the Second International Conference on Sensor Technologies and Application, 2008.

[20] Y. Tobe and T. Suzuki, "WISER: Cooperative Sensing using Mobile Robots" in the Proceedings of the 11th International Conference on Parallel and Distributed Systems, 2005.

[21] L. Parker, "ALLIANCE: An Architecture for Fault Tolerant Multi-Robot Cooperation", IEEE Transactions on Robotics and Automation, Vol. 14, No. 2, April 1998, pp. 220–240.

[22] B. B. Choudhury, B. B. Biswal and B. B. Mishra, "Development of Optimal Strategies for Task Assignment in Multi-robot Systems" in the Proceedings of the IEEE International Advanced Computing Conference, 2009.

[23] J. Albaric and R. Zapata, "Motion Planning of Cooperative Non-holonomic Mobile Manipulators" in the IEEE International Conference on Systems, Man & Cybernetics, 2002.

[24] Y. Wang and M. Tan, "A Multi-Robot Coordination System for Manipulation of Huge Elliptical Workpiece" in the Proceedings of the IEEE International Conference on Intelligent Systems and Signal Processing, 2003.

[25] G. A. S. Periera, B. S. Pimentel, L. Chaimowicz and M. F. M. Campos, "Coordination of Multiple Mobile Robots in an Object Carrying Task using Implicit Communication" in the Proceedings of the IEEE International Conference on Robotics and Automation, 2002.

[26] A. Trebi-Ollenu, H. D. Nayar, H. Aghazarian, A. Ganino, P. Pirjanian, B. Kennedy, T Huntsberger and P. Schenker, "Mars Rover Pair Cooperatively Transporting a Payload" in the Proceedings of the IEEE International Conference on Robotics and Automation, 2002.

[27] A. K. Bouloubasis and G. T. McKee, "Cooperative Transport of Extended Payloads" in the Proceedings of ICAR 2005, Seattle, pp. 882–887, 2005.

[28] M. T. Khan and C. W. de Silva, "Autonomous Fault Tolerant Multi-Robot Coordination for Object Transportation based on Artificial Immune System" in the Proceedings of the 2nd International Conference on Robot Communication and Coordination, Odense, Denmark, 2009.

[29] P. S. Schenker, T. L. Huntsberger, P. Pirjanian, E. Baumgartner, H.Aghazarian, A. Trebi-Ollennu, P. C. Leger, Y. Cheng, P. G. Backes, E. W. Tunstel, S. Dubowsky, K. Iagnemma, G. T. McKee, "obotic automation for space: planetary surface exploration, terrain-adaptive mobility, and multi-robot cooperative tasks," in Proc. SPIE Vol. 4572, Intelligent Robots and Computer Vision XX, Newton, MA, October, 2001.

[30] B. Varghese and G. T. McKee, "A Mathematical Model, Implementation and Study of a Swarm Conglomerate and its Formation Control" in the Proceedings of Towards Autonomous Robotic Systems, Edinburgh, Scotland, 2008, pp. 156–162.

[31] B. Varghese and G. T. McKee, "A Mathematical Model, Implementation and Study of a Swarm System" in the Robotics and Autonomous Systems Journal, Special Issue: Towards Autonomous Robotic Systems 2009: Intelligent, Autonomous Robotics in the UK', Vol.58, Issue 3, March 2010, pp.287–294.

[32] B. Varghese and G. T. McKee, "Towards a Unifying Mathematical Framework for Pattern Transformation in Swarm Systems" accepted for publication in the International Journal of Vehicle Autonomous Systems, Special Issue on "Modelling and/or Control of Multi-Vehicle Formations", 2010.

[33] MissionLab website: http://www.cc.gatech.edu/ai/robot-lab/research/MissionLab/

[34] MissionLab User Manual for MissionLab Version 7.0, Georgia Tech Mobile Robot Laboratory, College of Computing, Georgia Institute of Computing, Atlanta, GA 30332, July 12, 2006.

[35] B. Varghese and G. T. McKee, "A Swarm Pattern Transformation method based on Macroscopic Parameter Operation" in the Proceedings of the IEEE International Conference on Robotics and Biomimetics, Bangkok, Thailand, 2008, pp. 1164–1169.

[36] B. Varghese and G. T. McKee, "Modelling and Simulating a Mathematical Tool for Multi-robot Pattern Transformation" in the Proceedings of the International Conference on Computer Modelling and Simulation, Macau, China, 2009, pp. 21–27.

[37] M. J. Quinn, "Parallel Computing Theory and Practice" McGraw-Hill, Inc. 1994.

[38] J. P. Walters and V. Chaudhary, "Replication-Based Fault Tolerance for MPI Applications" in the IEEE Transactions on Parallel and Distributed Systems, Vol. 20, No. 7, July 2009, pp. 997–1010.

[39] X. Yang, Y. Du, P. Wang, H. Fu and J. Jia, "FTPA: Supporting Fault-Tolerant Parallel Computing through Parallel Recomputing" in the IEEE Transactions on Parallel and Distributed Systems, Vol. 20, Issue 10, October 2009, pp. 1471–1486.

[40] B. Varghese and G. T. McKee, "Can Space Applications Benefit from Intelligent Agents?" in the Proceedings of the 3rd International ICST Conference on Autonomic Computing and Communication Systems, Limassol, Cyprus, 2009 and in the Proceedings of AUTONOMICS 2009, LNICST 23, 2009, pp. 197–207.

[41] B. Varghese, G.T. McKee and V. N. Alexandrov, "A Cluster-Based Implementation of a Fault-Tolerant Parallel Reduction Algorithm Using Swarm-Array Computing" in the Proceedings of the 6th International Conference on Autonomic and Autonomous Systems, Cancun, Mexico, 2010, pp. 30–36.

[42] W. Gropp, E. Lusk and A. Skjullum, "Using MPI-2: Advanced Features of the Message Passing Interface", MIT Press (1999).

[43] MPI Tutorial: http://www.mpi-forum.org/docs/mpi-20-html/mpi2-report.html

[44] OpenMPI website: http://www.open-mpi.org/

[45] E. Gabriel, G. E. Fagg, G. Bosilca, T. Angskun, J. Dongarra, J. M. Squyres, V. Sahay, P. Kambadur, B. Barrett, A. Lumsdaine, R. H. Castain, D. J. Daniel, R. L. Graham, T. S. Woodall, "Open MPI: Goals, Concept, and Design of a Next Generation MPI Implementation" in the Proceedings of the 11th European PVM/MPI Users' Group Meeting, Budapest, Hungary, 2004, pp. 97–104.

[46] Adaptive MPI Manual Online: http://charm.cs.uiuc.edu/manuals/html/ampi/manual.html

[47] L. V. Kale and S. Krishnan, "CHARM++: Parallel Programming with Message-Driven Objects" in 'Parallel Programming using C++' (Eds. G. V. Wilson and P. Lu), pp. 175–213, MIT Press, 1996.

Send Orders of Reprints at reprints@benthamscience.net

CHAPTER 2

Collaborative Crowd Surveillance Using Networked Robotic Cameras[1]

Yiliang Xu and Dezhen Song

Department of Computer Science and Engineering

Texas A&M University

College Station, TX 77843, USA

---

[1]A preliminary version of this work has been published in Autonomous Robots (2010) 29: 53-66.

Nak Young Chong

ABSTRACT. We report an autonomous surveillance system with multiple networked pan-tilt-zoom (PTZ) cameras assisted by a fixed wide-angle camera. The wide-angle camera provides large but low resolution coverage and detects and tracks all moving objects in the scene. Based on the output of the wide-angle camera, the system generates spatiotemporal observation requests for each moving object, which are candidates for close-up views using PTZ cameras. Due to the fact that there are usually much more objects than the number of PTZ cameras, the system first assigns a subset of the requests/objects to each PTZ camera. The PTZ cameras then select the parameter settings that best satisfy the assigned competing requests to provide high resolution views of the moving objects. We propose an approximation algorithm to solve the request assignment and the camera parameter selection problems in real time. The effectiveness of the proposed system is validated in comparison with an existing work using simulation. The simulation results show that in heavy traffic scenarios, our algorithm increases the number of observed objects by over 210%.

1. Introduction

Consider a wide-angle networked camera is installed at an airport for human activity surveillance or in a forest for wildlife observation. The wide-angle camera can provide large, low resolution coverage of the scene. However, recognition and identification of humans and animals usually require close-up views at high resolution which needs PTZ cameras. The resulting autonomous observation system consists of a fixed wide-angle camera with multiple PTZ cameras as illustrated in Fig. 1. The wide-angle camera monitors the entire field to detect and track all moving objects. Each PTZ camera selectively covers a subset of the objects.

However there are usually more moving objects than the number of PTZ cameras. With these competing spatiotemporal observation requests, the major challenge is the control and scheduling of the PTZ cameras to maximize the "satisfaction" to the competing requests. The system design emphasizes the "satisfaction" to the requests which takes into account 1) camera coverage over objects, 2) camera zoom level selection, and 3) camera traveling time. We approach the control and scheduling problem in two steps. First, a subset of the requests/objects is assigned to each PTZ camera. Second, each PTZ camera selects its PTZ parameters to cover the assigned objects. We formulate the problems in both steps as frame selection problems and propose an approximation algorithm to solve them in real time. We implement the system and validate the system and the algorithm in simulations and physical experiments. The experimental results show that our method outperforms an existing work by increasing the number of observed objects by over 210% in heavy traffic scenarios.

2. Related Work

In the recent decade, multiple camera surveillance systems, especially those with both static and active cameras [1] have attracted growing attention of research. Most of the works are master-slave camera configuration [2]. The master static camera(s) provide the general information about the wide-angle scene while the slave active cameras acquire the localized high-resolution imagery of the regions of interest. This is a relatively new research area with many directions to explore. Our work belongs to this category.

Most works in this category schedule the active cameras based on straightforward heuristic rules. Zhou et al. [2] choose the object closest to the current camera setting as the next observation object. Hampapur et al. [3] adopt the simple round robin sampling.

Bodor et al. [4] and Fiore et al. [5] propose a dual-camera system with one wide-angle static camera and a PTZ camera for pedestrian surveillance. Human activities (walking, running, etc.) are prioritized based on the preliminary recognition by the wide-angle camera. The PTZ camera focuses to the activity with the highest priority for further analysis. Costello et al. [6] are the first to formulate the single camera scheduling problem based on network packet scheduling methods. The authors propose and compare several greedy scheduling policies. With different assumptions towards the observation scene and objects, various scheduling formulation and schemes are proposed. In Lim et al. [7], the scheduling problem is formulated as a graph matching problem. In Bimbo and Pernici [8], the continuous scheduling problem is truncated by a predefined observation deadline and each truncated camera scheduling problem is formulated as an online dynamic vehicle routing problem (DVRP). Qureshi and Terzopoulos [9] propose a virtual environment simulator to test various camera senor network frameworks. However these methods assign only one object to one active camera. Our system assigns multiple objects to individual cameras by selecting PTZ camera parameters such that the camera coverage-resolution tradeoff is achieved. This also enables group watching with scalability.

Very few work considers the selection of the zoom level of active cameras and assigns multiple objects to individual cameras. Lim et al. [10, 11] construct the observation task for each single object as a "task visibility interval" (TVI) based on its predicted states and corresponding camera settings. When TVIs have non-empty intersection, they are grouped to form a "multiple task visibility interval" (MTVI). Based on the order of the starting time of (M)TVIs, a directed acyclic graph (DAG) is constructed. The scheduling problem is formulated as a maximal flow problem. A greedy algorithm and a dynamic programming scheme are proposed to solve it. Zhang et al. [12] construct a semantic saliency map to indicate the observation requests. An exhaustive algorithm finds the optimal single frame that minimizes the information loss. Sommerlade and Reid [13] use an information-theoretic framework to study how to select a single active camera's zoom level for tracking a single object to balance the chances of loosing the tracked object and that of loosing trace of other objects. In contrast to these works, our approach dose not require accurate long-term motion prediction. The assignment of multiple objects to individual PTZ cameras is carried out by selecting the camera parameters to achieve the tradeoff between coverage and resolution.

The p-frame problem is structurally similar to the p-center facility location problem, which has been proven to be NP-complete [14]. Given n request points on a plane, the task is to optimally allocate p points as service centers to minimize the maximum distance (called min-max version) between any request point and its corresponding service center. In [15], an $O(n \log^2 n)$ algorithm for a 2-center problem is proposed. As an extension, replacing service points by orthogonal boxes, Arkin et al. [16] propose a $(1 + \epsilon)$-approximation algorithm that runs in $O(n \min(\lg n, 1/\epsilon) + (\lg n)/\epsilon^2)$ for the 2-box covering problem. Alt et al. [17] proposed a $(1 + \epsilon)$-approximation algorithm that runs in $O(n^{O(m)})$, where $\epsilon = O(1/m)$, for the multiple disk covering problem. The requests in these problems are all points instead of polygonal regions as those in our p-frame problem and the objective of the p-frame problem is to maximize the satisfaction, which is not a distance metric.

Our group focuses on developing intelligent vision systems and algorithms using robotic cameras for a variety of applications such as construction monitoring, distance learning, panorama construction and natural observation [18]. In the context of using PTZ camera for the collaborative observation, competing observation requests need to be covered by camera frame(s) to maximize the overall observation reward. This issue is formulated as a single frame selection (SFS) problem [**19**]. A series of algorithms for the single frame selection problem have been proposed [**19**, **20**]. Song et al. [**21**] propose an autonomous observation system in which a single PTZ camera is used to fulfill competing spatiotemporal observation requests. Here, multiple PTZ cameras are used to increase the observation coverage. We formulate the problem of coordinating the p camera frames as the p-frame problem and propose an approximation algorithm for solving it. This algorithm inspires the direction of simultaneous multi-object observation using multiple PTZ cameras as in this work.

3. System Architecture and Hardware

Fig. 1 shows the architecture of the system. The system consists of p $(p \geq 1)$ PTZ cameras and a wide-angle camera. All cameras are calibrated. The wide-angle camera detects and labels all moving objects in the scene. The states of the objects (e.g., size, position and velocity) are tracked and predicted. Based on the prediction, the observation request generation module generates the competing spatiotemporal observation requests (shadowed rectangles) for all objects. Then the request assignment module groups requests and assigns a subset of the objects/requests to each PTZ camera by computing the p-frame settings that best satisfy the requests. Each PTZ camera tracks the objects assigned to it by selecting the PTZ parameter settings that best satisfy these requests to capture high resolution images/videos of the objects.

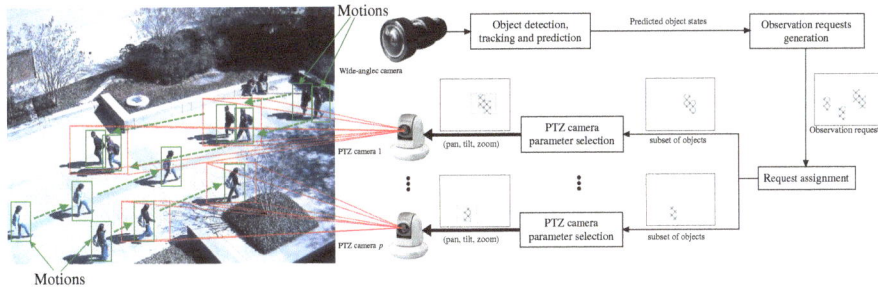

FIGURE 1. System architecture. The solid green rectangles represent the moving objects and the dashed red rectangles indicate the selective coverage of the PTZ cameras.

Fig. 2 shows the timeline of the system. An observation cycle starts at time $t = t_0$. The states of the objects at time $t = t_0 + \delta_l$ are predicted, where δ_l is termed as "lead time". Based on the predicted states, the system generates the observation request at time $t = t_0 + \delta_l$ for each object. A subset of these objects is then assigned to each PTZ camera. Then the system starts to adjust the PTZ cameras according to the request assignment. The camera traveling time is bounded below the "lead time" δ_l so that the cameras can intercept the objects at time $t = t_0 + \delta_l$. After that, each PTZ camera tracks its object subset for time δ_r until the beginning of the next observation cycle. δ_r is termed as "recording time" and is evenly divided into n_r intervals with each of length τ. Based on the state prediction, the PTZ camera parameter selection module computes each camera's setting at the end of each interval. Then each camera micro-adjusts its settings for up to τ time and prepares for the next interval. By capturing images\videos for δ_r time, the request assignment module re-initiates and the operations above repeat. $T = \delta_l + \delta_r$ is called one observation cycle.

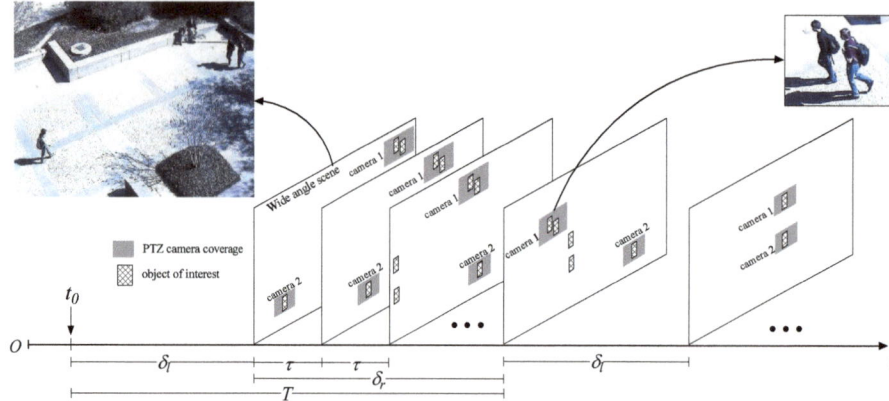

FIGURE 2. System timeline. An observation cycle starts at $t = t_0$. Within each cycle time $T = \delta_l + \delta_r$, all PTZ cameras first take no more than δ_l time to adjust the PTZ parameters so that each PTZ camera is assigned with a subset of the objects. Then each PTZ camera micro-adjusts its parameters within interval τ to track the assigned subset of objects. This tracking lasts δ_r time until a new observation cycle starts.

The stationary camera we use is a Arecont Vision AV3100 with a Computar lens whose focal length ranges from 4.5mm to 12.5mm. The camera runs at 11 fps with high resolution of 1600×1200. The PTZ cameras we use are Panasonic HMC280. The camera uses the MPEG4 compression and runs at up to 30 fps with resolution of 640×480. It has a $350°$ pan range and a $120°$ tilt range. It can pan and tilt up to $300°/s$ and $200°/s$, respectively. It has $21\times$ motorized zoom with zoom-changing speed up to 5 $level/s$ The system is programmed using Microsoft Visual C++.

4. Camera Scheduling

For p PTZ cameras, there are usually much more objects/requests. With the competing spatiotemporal requests, we need to control and schedule the PTZ cameras to capture sequences of images/videos that best satisfy the requests.

Observation Request Generation. The wide-angle camera detects moving objects and tracks them continuously. Each object is represented by its minimal iso-oriented bounding rectangular region which is determined by a 4-parameter vector,

$$(1) \qquad\qquad [u, v, a, b]^T,$$

where (u, v) indicates the center of the rectangle in the image space; a and b denote the width and height of the rectangle, respectively. Thus the state of the object at time t can be represented by

$$x(t) = [u(t), v(t), a(t), b(t), \dot{u}(t), \dot{v}(t)]^T,$$

where $(\dot{u}(t), \dot{v}(t))$ indicates the velocity of the rectangle center in the image space at time t.

A non-parametric Gaussian background subtraction model [**22**] is used to detect and label any moving objects. For tracking and predicting the object state, each labeled object is assigned a Kalman filter. A commonly used constant velocity model is adopted. Kalman filter is also able to handle short-term occlusion by predicting the object motion. It is worth mentioning that a lot of other existing tracking algorithms [**23**] can be applied here and the tracking itself is not the focus of this work.

Given the predicted state of i-th object at time t is

$$\hat{x}_i(t) = [\hat{u}_i(t), \hat{v}_i(t), \hat{a}_i(t), \hat{b}_i(t), \hat{\dot{u}}_i(t), \hat{\dot{v}}_i(t)]^T,$$

we define the spatiotemporal observation request of i-th object at time t as,

$$(2) \qquad\qquad r_i(t) = [T_i(t), \hat{b}_i(t), z_i, \omega_i(t)]^T,$$

where $T_i(t)$ represents the polygonal request region which is circumscribed by the rectangle determined by $\hat{u}(t)$, $\hat{v}(t)$, $\hat{a}_i(t)$ and $\hat{b}_i(t)$ in the same way as u, v, a and b in (1); z_i indicates the desirable resolution. $\omega_i(t)$ is the temporal weight, which indicates the emergency/importance level of the i-th object at time t. $\omega_i(t)$ plays an important role in balancing the observation service across all the objects and will be discussed in details later.

Request Assignment. As shown in Fig. 2, at the beginning of each recording time δ_r, we need to coordinate p PTZ cameras so that each camera is assigned a subset of the objects. We choose the p-frame settings that best satisfy all the requests at that time. Here we formulate it as an optimization problem termed p-frame problem.

Input and Output. The input of the p-frame problem is a set of n requests $R(t) = \{r_i(t)|i = 1, 2, ..., n\}$. Each request is defined as $r_i(t) = [T_i(t), z_i, \omega_i]$, where $T_i(t)$ denotes the polygonal requested region which is circumscribed by the rectangle determined by $[u(t), v(t), a(t), b(t)]^T$; $z_i \in Z$ specifies the desired resolution level, which is in the range of $Z = [\underline{z}, \overline{z}]$. The only requirement for $T_i(t)$ is that its coverage area can be computed in constant time.

A solution to the p-frame problem is a set of p PTZ camera frames. Given a fixed aspect ratio (e.g. 4:3), a camera frame can be defined as $c = [x, y, z]^T$, where pair (x, y) denotes the center point of the rectangular frame and $z \in Z$ specifies the resolution level of the camera frame. Here we consider the coverage of the camera as rectangular according to the camera configuration space. Therefore, the width and height of the camera frame can be represented as $4z$ and $3z$ respectively. The coverage area of the frame is $12z^2$. The four corners of the frame are located at $(x \pm \frac{4z}{2}, y \pm \frac{3z}{2})$.

Given w and h are the camera pan-tilt ranges respectively, then

$$\mathbb{C} = [0, w] \times [0, h] \times Z$$

defines the set of all candidate frames. Therefore, $\mathbb{C}^p$ indicates the solution space for the p-frame problem. We define any candidate solution to the p-frame problem as $C^p = (c_1, c_2, ..., c_p) \in \mathbb{C}^p$, where $c_i, i = 1, 2, ..., p$, indicates the i-th camera frame in the solution. In the rest of the chapter, we use superscription * to indicate the optimal solution. The objective of the p-frame problem is to find the optimal solution $C^{p*} = (c_1^*, c_2^*, ..., c_p^*) \in \mathbb{C}^p$ that best satisfies the requests.

Set Operators. We clarify the use of set operators such as "$\cap$", "$\subseteq$" and "$\nsubseteq$" to represent the relationship between frames, frame sets, and requests in the rest of the chapter.

- When two operands are frames or requests (e.g., $r_i(t) \in R(t)$, $c_u, c_v \in \mathbb{C}$), the set operators represent the 2-D regional relationship between them. For example,

$r_i(t) \subseteq c_u$ represents that the region of $r_i(t)$ is fully contained in that of frame c_u while $c_u \cap c_v$ represents the overlapping region of frames c_u and c_v.

- When the operands are one frame (e.g., $c_i \in \mathbb{C}$) and one frame set (e.g., $C^k \in \mathbb{C}^k, k < p$), we treat the frame as an element of a frame set. For example, $c_i \notin C^k$ represents that c_i is not an element frame in the frame set C^k.

- When the operands are two frame sets, we use set operators. For example, $\{c_1\} \subset C^p$ means frame set $\{c_1\}$ is a subset of C^p. Frame set $\{c_1, c_2\} = \{c_1\} \cup \{c_2\}$ is different from $c_1 \cup c_2$. The former is the frame set that consists of two element frames and the later is the union area of the two frames.

Assumptions. We assume that the p frames are taken from p cameras that share the same workspace. Therefore, if a location can be covered by a camera frame, other frames can cover that location, too.

We assume that the solution C^{p*} to the p-frame problem satisfies the following condition.

DEFINITION 1 (Least Overlapping Condition (LOC)). $\forall r_i(t), i = 1, ...n, \forall c_u \in C^{p*}$, $\forall c_v \in C^{p*}$, and $c_u \neq c_v$,

$$(3) \qquad\qquad r_i(t) \nsubseteq c_u \cap c_v.$$

The LOC means that the overlap between camera frames is so small that no request can be fully covered by more than one frame simultaneously. The LOC forces the overall coverage of a p-frame set $\cup_{j=1}^{p} c_j$ to be close to the maximum. This is meaningful in applications when the cameras need to search for unexpected events while best satisfying the n existing requests because the ability to search is usually proportional to the union of overall coverage. Therefore, the LOC can increase the capability of searching for unexpected events. The extreme case of the LOC is that there is no overlap between camera frames.

DEFINITION 2 (Non-Overlapping Condition (NOC)). *Given a p-frame set $C^p = (c_1, c_2, ..., c_p) \in \mathbb{C}^p$ ($p \geq 2$), C^p satisfies the NOC, if*

$$\forall u = 1, 2, ..., p, \forall v = 1, 2, ..., p, u \neq v, c_u \cap c_v = \phi .$$

It is not difficult to find that the NOC is a sufficient condition to the LOC. The NOC yields the maximum union coverage and is a favorable solution to applications where searching ability is important.

Satisfaction Metric. To measure how well a p-frame set satisfies the requests, we need to define a satisfaction metric. We extend the Coverage-Resolution Ratio (CRR) metric in [**19**] and propose a new Resolution Ratio with Non-Partial Coverage (RRNPC).

DEFINITION 3 (RRNPC metric). *Given a request $r_i(t) = [T_i(t), z_i, \omega_i(t)]$ and a camera frame $c = [x, y, z]^T$, the satisfaction of request $r_i(t)$ with respect to c is computed as*

$$(4) \qquad\qquad s(c, r_i(t)) = \omega_i(t) \cdot I(c, r_i(t)) \cdot \min(\frac{z_i}{z}, 1),$$

where $I(c, r_i(t))$ is an indicator function that describes the non-partial coverage condition,

$$(5) \qquad\qquad I(c, r_i(t)) = \begin{cases} 1 & if \;\; r_i(t) \subseteq c, \\ 0 & otherwise. \end{cases}$$

Eq. (5) indicates that we do not accept partial coverage over the request. Only the requests completely contained in a camera frame contribute to the overall satisfaction. From (4) and (5), the satisfaction of the ith request is a scalar $s_i \in [0, 1]$.

Based on (4), the satisfaction of $r_i(t)$ with respect to a candidate least overlapping p-frame set $C^p = (c_1, c_2, ..., c_p) \in \mathbb{C}^p$ is,

$$(6) \qquad s(C^p, r_i(t)) = \sum_{u=1}^{p} I(c_u, r_i(t)) \cdot \min(\frac{z_i}{z_u}, 1),$$

where z_i, z_u indicate the resolution values of $r_i(t)$ and the u-th camera frame in C^p respectively. The LOC implies that although (6) is in the form of summation, at most one frame contains the region of request $r_i(t)$ and thus non-negative $s(C^p, r_i(t))$ has a maximum value of 1. Therefore, RRNPC is a standardized metric that takes both the region coverage and the resolution level into account.

To simplify the notation, we use $s(c) = \sum_{i=1}^{n} s(c, r_i(t))$ to represent the overall satisfaction of a single frame c. We also use $s(C^k) = \sum_{j=1, c_j \in C^k}^{k} s(c_j)$, to represent the overall satisfaction of a partial candidate k-frame set C^k, $k \le p$.

Problem Formulation. Based on the assumption and the RRNPC metric definition above, the overall satisfaction of a p-frame set $C^p = \{c_1, c_2, ..., c_p\} \in \mathbb{C}^p$ over n requests is the sum of the satisfaction of each individual request $r_i(t), i = 1, 2, ..., n$,

$$(7) \qquad s(C^p) = \sum_{i=1}^{n} \sum_{u=1}^{p} I(c_u, r_i(t)) \cdot \min(\frac{z_i}{z_u}, 1).$$

Eq. (7) shows that the satisfaction of any candidate C^p can be computed in $O(pn)$ time. Now we can formulate the least overlapping p-frame problem as a maximization problem,

$$(8) \qquad C^{p*} = \arg \max_{C^p \in \mathbb{C}^p} s(C^p).$$

To solve (8), we propose an approximation algorithm which will be presented later in Section 5. After assigning the requests by finding the optimal p frame settings, we find the best camera-setting pairs that minimize the time for adjusting the PTZ cameras.

PTZ Camera Parameter Selection. After each camera is assigned a subset of objects, the camera tries to track these objects for the recording time δ_r. This requires to select the camera parameter setting such that the satisfaction is maximized for each recording interval. Given each recording interval is represented as $[t - \tau, t]$ and the i-th camera is assigned a subset of objects with predicted states at time t, $\hat{X}_i(t) = \{\hat{x}_1(t), \hat{x}_2(t), ...\}$. The corresponding observation requests are generated $R_i(t) = \{r_1(t), r_2(t), ...\}$. The camera setting at time t, $c^*(t)$, is then determined by maximizing the satisfaction to $R_i(t)$,

$$(9) \qquad c^*(t) = \arg \max_{c} \sum_{r_i(t) \in R_i(t)} s(c, r_i(t)).$$

This problem can be solved using the approximation algorithm in [20] with running time $O(|\hat{X}_i|/\epsilon^3)$, where $|\hat{X}_i|$ is the cardinality of $\hat{X}_i$ and ϵ is the approximation bound. However, (9) does not consider the fact that within time τ, the PTZ camera can only microadjust within a limited setting range. We assume the pan, tilt and zoom motion of the camera are independent. The reachable ranges for pan, tilt and zoom settings within time τ are α, β and γ, respectively. Then we rewrite (9) as,

$$(10) \qquad c^*(t) = \arg \max_{c \in \alpha \times \beta \times \gamma} \sum_{r_i(t) \in R_i(t)} s(c, r_i(t)).$$

An interesting case is that the assigned objects scatter away in different directions. This case does not pose any trouble to our algorithm at all. As shown in (10), during each step of τ, the camera parameter selection module searches for the optimal camera setting within the reachable range $[\alpha \times \beta \times \gamma]$ that maximizes the satisfaction. It is possible to lose coverage over certain objects, but it is guaranteed that camera PTZ setting is optimal with respect to the object subset assigned to the camera.

It is worth mentioning that most PTZ cameras' pan and tilt motion is fast enough to follow most objects in the scene. For example, recall the transition speed of the Panasonic HCM 280 camera is $300°/s$ for pan, $200°/s$ for tilt and 5 $level/s$ for zoom, respectively. Considering the camera has $21\times$ zoom levels and only less than $50°$ FOV, the time for changing pan and tilt settings is much less than the time for changing the camera zoom. Changing the zoom level when the camera is moving also creates significant motion blurring and requires re-focusing. Therefore, in practice, we can search for the pan and tilt settings in $\alpha \times \beta$ while maintain the same zoom level for each recording period.

Dynamic Weighting. If we keep the request weight in (2) unchanged, the system will create a "biased frame selection" model that always prefers certain objects instead of balancing the camera resource for all objects. We address this issue by carefully designing the temporal weight $\omega_i(t)$ based on two intuitions: 1) object exiting FOV sooner is of more importance and 2) object less satisfied in history is of more importance. The first intuition is derived from the earliest deadline first (EDF) policy [6]. The policy addresses the emergency of the requests. The second intuition addresses sharing the camera resource for all objects to achieve balanced observation over time. We define,

$$\omega_i(t) = \mu_i(t) \cdot \nu_i(t),$$

where $\mu_i(t)$ and $\nu_i(t)$ address the first and second intuitions, respectively. One candidate form of $\mu_i(t)$ is,

$$(11) \qquad \mu_i(t) = \min(\rho^{(\hat{d}_i - t)}, 1),$$

where $\hat{d}_i$ is the predicted deadline for i-th object to exit the FOV and $0 < \rho < 1$ is a parameter that controls how quick the emergency increases. When $t \to \hat{d}_i$, $\mu_i(t) \to 1$, as maximum.

To design $\nu_i(t)$ we need to first define the accumulative unweighted satisfaction (AUS) $\eta_i(t)$,

$$(12) \qquad \eta_i(t) = \sum_{j=1}^{p} \sum_{t_k \leq t} \frac{s(c_j(t_k), r_i(t_k))}{\omega_i(t_k)},$$

where the variable t_k refers to the discrete times when cameras take frames. The AUS essentially reflects how well an object is satisfied in history. We design $\nu_i(t)$ as,

$$(13) \qquad \nu_i(t) = \max(1 - \frac{\eta_i(t)}{n_e}, 0),$$

where n_e is a parameter indicating the extent to which an object need to be observed. When $\eta_i(t) \geq n_e$, $\nu_i(t)$ is zero and we contend the object is fully satisfied and needs no observation any longer. Both $\mu_i(t)$ and $\nu_i(t)$ are bounded in range $[0, 1]$, which keeps the satisfaction metric in (4) a standard metric.

It is worth mentioning that the time for computing the weights $\omega_i(t)$ is almost instant. Eq. (11) takes instant time. The AUS as in (12) can be computed in an incremental way as the observation progresses. Therefore, the overall time for generating the observation requests is negligible.

5. p-Frame Selection Algorithm

Solving the optimization problem in (8) is nontrivial. To enumerate all possible combinations of candidate solutions by brute force can easily take up to $O(n^p)$ time. In this section, we present a lattice-based approximation algorithm beginning with the construction of the lattice. To maintain the LOC in the lattice framework, we introduce the Virtual Non-Overlapping Condition(VNOC). Based on the VNOC, we analyze the structure of the approximate solution and derive the approximation bound with respect to the optimal solution that satisfies the NOC. To summarize this, a lattice-based induction-like algorithm is presented at the end of the section.

Construction of Lattice. We construct a regular 3-D lattice, which is inherited from [20] to discretize the solution space $\mathbb{C}^p$. Let 2-D point set $V = \{(\alpha d, \beta d) | \alpha d \in [0, w], \beta d \in [0, h], \alpha, \beta \in \mathcal{N}\}$ discretize the 2-D reachable region and represent all candidate center points of rectangular frames, where d is the spacing of the pan and tilt samples. Let 1-D point set $\mathcal{Z} = \{\gamma d_z | \gamma d_z \in [\underline{z}, \overline{z} + 2d_z], \gamma \in \mathcal{N}\}$ discretize the feasible resolution range and represent all candidate resolution values for the camera, where d_z is the spacing of the zoom. Therefore, we can construct the lattice as a set of 3-D points, $L = V \times \mathcal{Z}$.

Each point $c = (\alpha d, \beta d, \gamma d_z) \in L$ represents the setting of a candidate camera frame. There are totally $(wh/d^2)(g/d_z) = |L|$ candidate points/frames in L, where $g = \overline{z} - \underline{z}$. We set $d_z = d/3$ for cameras with an aspect ration of $4 : 3$ according to [20].

What is new is that the spacing of the lattice d and d_z also depends on the size of the requested regions. For any request $r_i \in R$, there exists an Iso-oriented Bounding Box (IBB) for each r_i. Let us define λ and μ as the smallest width and height across all IBBs, respectively. We choose d such that

$$(14) \qquad\qquad d < \min(3\lambda/10, \mu/3).$$

This input-sensitive lattice setting can help us to establish the LOC on the lattice and will be discussed later in Section 5. From here on, we use symbol $\tilde{\ }$ to denote the lattice-based notations. For example, $\tilde{C}^p$ denotes a p-frame set on lattice L.

DEFINITION 4. *For any camera frame* $c \in \mathbb{C}$,

$$\tilde{c}' = \min \tilde{c}, \ s.t. \ \tilde{c} \in L \ and \ c \subseteq \tilde{c}.$$

Hence $\tilde{c}'$ *is the smallest frame on the lattice that fully encloses c.*

In the rest of the chapter, we use symbol $'$ to denote the corresponding smallest frame(s) on the lattice. For any camera frame $c = [x, y, z]$ and its corresponding $\tilde{c}' = [\tilde{x}', \tilde{y}', \tilde{z}']$, we define their bottom-left corners as (x^l, y^b) and $(\tilde{x}'^l, \tilde{y}'^b)$, and their top-right corners as (x^r, y^t) and $(\tilde{x}'^r, \tilde{y}'^t)$, respectively.

From the results of [20], we have

$$(15) \qquad \begin{aligned} x^l - \tilde{x}'^l \leq 5d/3, \quad \tilde{x}'^r - x^r \leq 5d/3, \\ y^b - \tilde{y}'^b \leq 3d/2, \quad \tilde{y}'^t - y^t \leq 3d/2. \end{aligned}$$

Virtual Non-Overlapping Condition. The NOC defined in Definition 2 guarantees the LOC. However, due to the limitation of lattice spacing, it is very difficult for candidate frames on the lattice to follow the NOC. Actually, it is unnecessary (though sufficient) to follow the NOC to satisfy the LOC. It is possible to allow a minimum overlap that is controlled by the lattice spacing and meanwhile guarantee that the LOC is still satisfied, which yields the Virtual Non-Overlapping Condition (VNOC).

DEFINITION 5 (Virtual Non-Overlapping Condition(VNOC)). *Given any j-frame set $C^j = (c_1, c_2, ..., c_j) \in \mathbb{C}^j, j = 2, 3, ..., p$ and any two frames $c_u, c_v \in C^j$, then C^j satisfies the VNOC, if*

$$\min(x_u^r - x_v^l, x_v^r - x_u^l) \leq 10d/3 \ \ or \ \ \min(y_u^t - y_v^b, y_v^t - y_u^b) \leq 3d.$$

COROLLARY 1. *Given any two frames $c_1, c_2 \in \mathbb{C}$, if $\{c_1, c_2\}$ satisfies the VNOC, then $\{c_1, c_2\}$ also satisfies the LOC.*

PROOF. From the definition of VNOC and the settings of λ and μ, we see that the size of the overlapping region $c_1 \cap c_2$, on either the x-axis or y-axis, is less than the size of the smallest request. This guarantees that no requested region is fully contained in the overlapping region. Therefore, the LOC is satisfied. $\square$

LEMMA 1. *Given any two frames $c_1, c_2 \in \mathbb{C}$ such that $\{c_1, c_2\}$ satisfies the VNOC, then*

$$(16) \qquad\qquad s(\{c_1, c_2\}) = s(c_1) + s(c_2).$$

PROOF. From Corollary 1, $\{c_1, c_2\}$ satisfies the LOC. From the definition of the LOC and the RRNPC satisfaction metric defined in (4), the conclusion follows. $\square$

Approximation Solution Bound. The construction of the lattice allows us to search for the best p frames on the lattice, which yields an approximation solution. Furthermore, the VNOC and Lemma 1 assist us in deriving the approximation bound.

LEMMA 2. *For any two frames $c_1, c_2 \in \mathbb{C}$, if $\{c_1, c_2\}$ satisfies the NOC, then $\{\tilde{c}_1', \tilde{c}_2'\}$ satisfies the VNOC.*

PROOF. The proof of the lemma is straightforward based on the definition of VNOC and the settings of λ and μ. $\square$

Given the optimal solution $C^{p*} = (c_1^*, c_2^*, ..., c_p^*)$ for the optimization problem defined in (8) that satisfies the NOC, there is a solution on the lattice $\tilde{C}'^{p*} = (\tilde{c}_1'^*, \tilde{c}_2'^*, ..., \tilde{c}_p'^*)$ whose element frames are the corresponding smallest frames on the lattice that contain those of C^{p*}. Lemma 2 implies that $\tilde{C}'^{p*}$ exists and satisfies the VNOC. However, how good is this solution in comparison to the optimal solution? Based on Theorem 1 in [20], we have

$$s(\tilde{c}_i'^*)/s(c_i^*) \geq 1 - \epsilon, \ i = 1, 2, ..., p,$$

where

$$(17) \qquad\qquad \epsilon = \frac{2d_z}{\underline{z} + 2d_z}.$$

Based on Lemma 1, we have

$$s(\tilde{C}'^{p*})/s(C^{p*}) \geq 1 - \epsilon.$$

where the approximation bound ϵ characterizes the comparative ratio of the approximation solution $\tilde{C}'^{p*}$ to the optimal solution C^{p*}. Eq. (17) indicates that when $\epsilon \to 0$,

$$(18) \qquad d = 3d_z = \frac{3}{2}(\frac{\epsilon}{1-\epsilon})\underline{z}.$$

Eqs. (17) and (18) imply that we can control the quality of the approximate solution by tuning the lattice spacing d. On the other hand, based on the lattice structure and the definition of the approximation bound, we know,

$$(19) \qquad |L| = O(1/\epsilon^3).$$

Lattice-based Algorithm. With the approximation bound established, the remaining task is to search $\tilde{C}^{p*}$ on L. We design an induction-like approach that builds on the relationship between the solution to the $(p-1)$-frame problem and the solution to the p-frame problem. The key elements that establish the connection are Conditional Optimal Solution (COS) and Conditional Optimal Residual Solution (CORS).

DEFINITION 6 (Conditional Optimal Solution). $\forall \tilde{c} \in L$, the COS, $\tilde{U}_j(\tilde{c}) = \{\tilde{C}^{j*}|\tilde{c} \in \tilde{C}^{j*}\}$, is defined as the optimal j-frame set, $j = 1, 2, ..., p$, for the j-frame problem that must include $\tilde{c}$ in the solution set. Also, $\tilde{U}_j(\tilde{c})$ satisfies the VNOC.

Therefore, we can obtain the optimal solution, $\tilde{C}^{p*}$, on the lattice by searching $\tilde{c}$ over L and its corresponding COS,

$$(20) \qquad \tilde{C}^{p*} = \tilde{U}_p(\tilde{c}^*),$$

where $\tilde{c}^* = \arg\max_{\tilde{c} \in L} s(\tilde{U}_p(\tilde{c}))$.

DEFINITION 7 (Conditional Optimal Residual Solution). *Given any COS*, $\tilde{U}_{j+1}(\tilde{c}), j = 0, 1, ..., p-1$, we define the j-frame CORS with respect to $\tilde{c}$ as: $\tilde{Q}_j(\tilde{c}) = \tilde{U}_{j+1}(\tilde{c}) - \{\tilde{c}\}$.

COROLLARY 2. $\tilde{Q}_j(\tilde{c})$ is the optimal j-frame set that satisfies,

- $\tilde{c} \notin \tilde{Q}_j(\tilde{c})$,
- $\{\tilde{c}\} \cup \tilde{Q}_j(\tilde{c})$ satisfies the VNOC.

What is interesting is that CORS allows us to establish the relationship between $\tilde{Q}_j$ and $\tilde{Q}_{j-1}$.

LEMMA 3.

$$(21) \qquad \tilde{Q}_j(\tilde{c}_u) = \tilde{Q}_{j-1}(\tilde{c}^*) \cup \{\tilde{c}^*\},$$

where $\tilde{c}^* = \arg\max_{\tilde{c} \in L} s(\tilde{Q}_{j-1}(\tilde{c}) \cup \{\tilde{c}\})$, subject to the constraint that $\{\tilde{c}_u, \tilde{c}\} \cup \tilde{Q}_{j-1}(\tilde{c})$ satisfies the VNOC.

PROOF. We prove the lemma by contradiction. Notice that the right hand side of (21) returns one of the j-frame sets that satisfy the two conditions in Corollary 2, while the left hand side is defined to be the optimal j-frame set that satisfies the same two conditions. Therefore, if we assume (21) does not hold, the only possibility is,

$$(22) \qquad s(\tilde{Q}_j(\tilde{c}_u)) > s(\tilde{Q}_{j-1}(\tilde{c}^*) \cup \{\tilde{c}^*\}).$$

Take an arbitrary frame $\tilde{c}_v \in \tilde{Q}_j(\tilde{c}_u)$ out of $\tilde{Q}_j(\tilde{c}_u)$, the result is $\tilde{Q}_j(\tilde{c}_u) - \{\tilde{c}_v\}$ and according to Lemma 1, we have,

$$(23) \qquad s(\tilde{Q}_j(\tilde{c}_u) - \{\tilde{c}_v\}) = s(\tilde{Q}_j(\tilde{c}_u)) - s(\tilde{c}_v).$$

Take $\tilde{c}_v$ out of $\tilde{Q}_{j-1}(\tilde{c}_v) \cup \{\tilde{c}_v\}$, the result is $\tilde{Q}_{j-1}(\tilde{c}_v)$ and

$$(24) \qquad\qquad s(\tilde{Q}_{j-1}(\tilde{c}_v)) = s(\tilde{Q}_{j-1}(\tilde{c}_v) \cup \{\tilde{c}_v\}) - s(\tilde{c}_v).$$

Based on (22) and the fact that

$$s(\tilde{Q}_{j-1}(\tilde{c}^*) \cup \{\tilde{c}^*\}) \geq s(\tilde{Q}_{j-1}(\tilde{c}_v) \cup \{\tilde{c}_v\}),$$

we have,

$$(25) \qquad\qquad s(\tilde{Q}_j(\tilde{c}_u)) > s(\tilde{Q}_{j-1}(\tilde{c}_v) \cup \{\tilde{c}_v\}).$$

Take $\tilde{c}_v$ out of both sides and combine with (23) and (24) respectively, we have,

$$(26) \qquad\qquad s(\tilde{Q}_j(\tilde{c}_u) - \{\tilde{c}_v\}) > s(\tilde{Q}_{j-1}(\tilde{c}_v)).$$

The frame set on the right hand side of (26), $\tilde{Q}_{j-1}(\tilde{c}_v)$, is defined to be the optimal $(j-1)$-frame set that satisfies the two conditions in Corollary 2 while the frame set on left hand side, $\tilde{Q}_j(\tilde{c}_u) - \{\tilde{c}_v\}$, is only one of the $(j-1)$-frame sets that satisfy the two conditions. Contradiction occurs. $\qquad\square$

It is worth mentioning that it takes $O(p)$ time to check if $(\{\tilde{c}_u, \tilde{c}\} \cup \tilde{Q}_j(\tilde{c}))$ satisfies the VNOC. Because $\{\tilde{c}\} \cup \tilde{Q}_j(\tilde{c}) = \tilde{U}_{j+1}(\tilde{c})$ satisfies the VNOC as defined in Definition 6 and thus we only need to check if $\{\tilde{c}_u\} \cup \tilde{U}_{j+1}(\tilde{c})$ satisfies the VNOC, which takes $O(p)$ time.

Eq. (20) implies that we can obtain the approximation solution $\tilde{C}^{p*}$ from $\tilde{U}_p$. Definition 7 indicates that we can obtain $\tilde{U}_p$ from $\tilde{Q}_{p-1}$. Now Lemma 3 implies that we can construct $\tilde{Q}_j$ from $\tilde{Q}_{j-1}, j = 1, 2, ..., p-1$. Considering the fact that $\tilde{Q}_0 = \phi$, this allows us to establish the algorithm using an induction-like approach. Algorithm 1 shows the complete lattice-based algorithm. Considering any candidate frame $\tilde{c} \in L$, we pre-calculate the satisfaction values for all the $|L|$ candidate frames and store the values in a lookup table to avoid redundant calculation. Given any candidate frame $\tilde{c}_u \in L$ as the input, the lookup function l returns the satisfaction value of $\tilde{c}_u$, $l(\tilde{c}_u) = s(\tilde{c}_u)$. We implement the lookup function using the array, $l[u] = s(\tilde{c}_u)$. From the pseudo code in Algorithm 1, it is not difficult to know that,

THEOREM 1. *Algorithm 1 runs in $O(n/\epsilon^3 + p^2/\epsilon^6)$ time.*

Speed up the algorithm. The algorithm introduced above is not fast enough for practical applications. Next we propose a brand and bound(BnB)-like approach and a lattice pruning (LP) scheme to improve the speed of the algorithm.

Branch and Bound-like Approach. Careful analysis reveals that the dominating factor of Algorithm 1 is the inner loop between line 20 and 26, which scans each node on lattice for updating the CORS. We found that it is not necessary to look at every node on L.

LEMMA 4. *Given any candidate frame $c = (x, y, z) \in L$, its (up to) nine child nodes on the lower layer (frames with zoom $z + d_z$)*

$$\begin{array}{lll} (x-d, y-d), & (x-d, y), & (x-d, y+d), \\ (x, y-d), & (x, y), & (x, y+d), \\ (x+d, y-d), & (x+d, y), & and\ (x+d, y+d) \end{array}$$

Algorithm 1: Lattice-based Algorithm

1	**begin**	
2	**for** $j \leftarrow 1$ **to** $|L|$ **do**	$O(1/\epsilon^3)$
3	$l[j] = s(\tilde{c}_j)$	$O(n)$
4	$\tilde{Q}_0(\tilde{c}_j) = \emptyset$;	$O(1)$
5	$s(\tilde{Q}_0(\tilde{c}_j)) = 0$;	$O(1)$
6	**end**	
7	**for** $k \leftarrow 1$ **to** p **do**	$O(p)$
8	$\tilde{C}^{k*} = \emptyset$;	$O(1)$
9	$s(\tilde{C}^{k*}) = 0$;	$O(1)$
10	**for** $u \leftarrow 1$ **to** $|L|$ **do**	update $\tilde{C}^{k*}, O(1/\epsilon^3)$
11	**if** $s(\tilde{C}^{k*}) < s(\tilde{Q}_{k-1}(\tilde{c}_u)) + l[u]$	
12	**then**	
13	$\tilde{C}^{k*} = \tilde{Q}_{k-1}(\tilde{c}_u) \cup \{\tilde{c}_u\}$;	$O(1)$
14	$s(\tilde{C}^{k*}) = s(\tilde{Q}_{k-1}(\tilde{c}_u)) + l[u]$;	$O(1)$
15	**end**	
16	**end**	
17	**for** $u \leftarrow 1$ **to** $|L|$ **do**	update $\tilde{Q}_k(\tilde{c}_u), O(1/\epsilon^3)$
18	$\tilde{Q}_k(\tilde{c}_u) = \tilde{Q}_{k-1}(\tilde{c}_u) \cup \emptyset$;	$O(1)$
19	$s(\tilde{Q}_k(\tilde{c}_u)) = s(\tilde{Q}_{k-1}(\tilde{c}_u))$;	$O(1)$
20	**for** $v \leftarrow 1$ **to** $|L|$ **do**	$O(1/\epsilon^3)$
21	**if** $s(\tilde{Q}_k(\tilde{c}_u)) < s(\tilde{Q}_{k-1}(\tilde{c}_v)) + l[v]$ **AND** $\{\tilde{c}_u, \tilde{c}_v\} \cup \tilde{Q}_{k-1}(\tilde{c}_v)$	
	satisfies the VNOC	$O(p)$
22	**then**	
23	$\tilde{Q}_k(\tilde{c}_u) = \tilde{Q}_{k-1}(\tilde{c}_v) \cup \{\tilde{c}_v\}$;	$O(1)$
24	$s(\tilde{Q}_k(\tilde{c}_u)) = s(\tilde{Q}_{k-1}(\tilde{c}_v)) + l[v]$;;	$O(1)$
25	**end**	
26	**end**	
27	**end**	
28	**end**	
29	**return** $\tilde{C}^{p*}$;	$O(1)$
30	**end**	

---

contain frame c.

The proof of this lemma is straightforward from the construction of the lattice L in Section 5. Lemma 4 suggests that if frame c fails the LOC check in line 21 of the Algorithm 1, its (up to) nine child nodes also fail the check because they all contain frame c and so do their child nodes. Fig. 3 illustrates this relationship. It is shown that frame c and its (up to) nine child nodes constitute a shape of pyramid.

COROLLARY 3. *Given any candidate frame $c = (x, y, z) \in L$ fails the LOC check and should be discarded, its (up to) nine child nodes can be discarded.*

Corollary 3 leads to a recursive scheme that can reduce the search space. Recall that we need to search through every nodes in the lattice to update the CORS. As shown in Algorithm 2, initially we label every node in lattice as "not discarded" and we search in increasing order of z. In the progress of the search, for any node either labelled as "discarded" or fail the LOC check, its (up to) nine child nodes are labelled as "discarded". This way we can reduce the search space significantly.

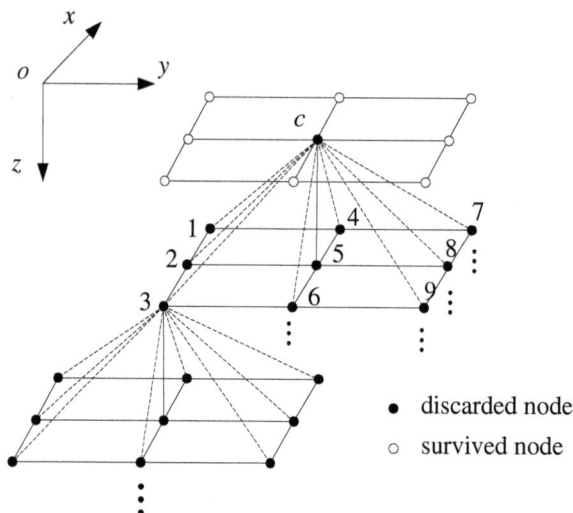

FIGURE 3. Illustration of the BnB-like approach.

Algorithm 2: BnB-like Approach	
1 **for** *each lattice node c=(x,y,z) in increasing order of z* **do**	$O(1/\epsilon^3)$
2 **if** *the node is labelled as "discarded"* **then**	
3 **if** *not the lowest layer* **then**	
4 label its (up to) nine child nodes as "discarded";	$O(1)$
5 **end**	
6 **else**	
7 **if** *c fails the LOC check* **then**	
8 **if** *not the lowest layer* **then**	
9 label its (up to) nine child nodes as "discarded";	$O(1)$
10 **end**	
11 **else**	
12 update CORS using c and $Q(c)$ as in Algorithm 1;	$O(1)$
13 **end**	
14 **end**	
15 **end**	

It is worth mentioning that, the BnB-like approach above is similar to but different from that in [20]. In [20], each lattice node only represents a candidate frame while in our work, each lattice node not only represents a candidate frame, but also corresponds to its CORS. The BnB-like approach in [20] rejects a lattice node by checking the resolution ratio and satisfaction ratio between two frames while our work rejects a lattice node by checking the LOC condition. As a result, in [20], once a lattice node is discarded, its (up to) 9 neighbor nodes with smaller resolution z, will be discarded. On the contrary, in our work, once a lattice node is discarded, its (up to) 9 neighbor nodes with larger resolution z, will be discarded. Therefore, we search in a z-increasing order while the work in [20] searches in a z-decreasing order. Readers are referred to [20] for details.

Lattice Pruning (LP). In Algorithm 1, the computation time is dominated by $|L|^2$. If we can reduce $|L|$, the computation time can be significantly improved. Close observation reveals that there are a lot of nodes on L that cannot contribute to the optimal solution.

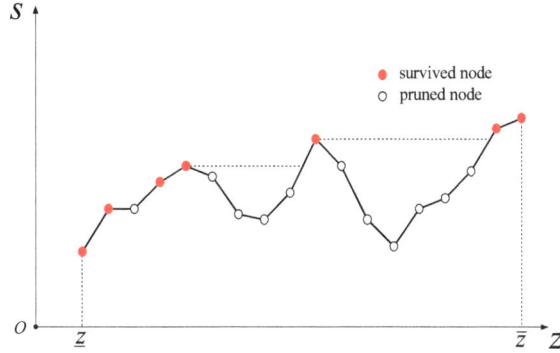

FIGURE 4. Lattice pruning for co-centered frames on L

LEMMA 5. *For any two co-centered candidate frame on L, $\tilde{c}_1 = [x, y, z_1]^T$ and $\tilde{c}_2 = [x, y, z_2]^T$, if $z_1 < z_2$ and $s(\tilde{c}_1) > s(\tilde{c}_2)$, then frame $\tilde{c}_2 \notin \tilde{C}^{p*}$ and thus can be discarded.*

PROOF. Since $\tilde{c}_1$, $\tilde{c}_2$ co-center and $z_1 < z_2$, we have $\tilde{c}_1 \subseteq \tilde{c}_2$. If $\tilde{c}_2 \in \tilde{C}^{p*}$, then $(\tilde{C}^{p*} \setminus \{\tilde{c}_2\} \cup \{\tilde{c}_1\})$ is a better p-frame set on L which satisfies the LOC. Contradiction occurs. □

Fig. 4 plots the satisfaction of a sample set of co-centered frames on L. It is shown that by searching from $\underline{z}$ up to $\bar{z}$, any frame whose satisfaction is less than that of the best frame previously searched can be discarded. This pruning can be applied during the initial calculation of the satisfaction of each candidate frame on L. This way, the search space is reduced significantly and the computation time is enhanced dramatically.

6. Experiment

We have implemented Algorithm 1 and the system using Microsoft Visual C++ 2005. The computer used is a Windows XP desktop PC with 2.0 GB RAM, 300 GB hard disk space and an Intel Pentium(R) Dual Core 3.2 GHz CPU. We first test the speed of Algorithm 1 using random inputs. Then we carry out a simulation to compare the camera scheduling of our system with an existing work based on the overall number of objects being observed. Finally, a physical experiment for crowd surveillance using real video data is reported.

Speed Test. In this experiment, we test the approximation algorithm speed with different parameter settings including the number of requests n, the number of camera frames p, and the approximation bound ϵ. In the experiment, both triangular and rectangular inputs are randomly generated. First, s_d points in V are uniformly generated across the reachable field of view. These points indicate the locations of interest and are referred to as seeds. Each seed is associated with a random radius of interest. To generate a request, we randomly assign it to one seed. For a triangular request, three 2-D points are randomly generated within the radius of the corresponding seed as the vertices of the triangle. For a rectangular request, a 2-D point is randomly generated as the center of the rectangular region within the radius of corresponding seed and then two random numbers are generated as the width and height of the request. Finally, the resolution value of the request is uniformly randomly generated across the resolution range $[\underline{z}, \overline{z}]$. Across the experiment, we set w=80, h=60, $\underline{z}$=5, $\overline{z}$=15 and s_d=4. For each parameter setting, 50 trials have been carried out for averaged performance.

We first test the speed performance of different implementations of the algorithm. Fig. (5) shows the relationship between the computation time and the number of frames p under three implementations: the original algorithm, the original algorithm with BnB approach (termed as BnB), and the original algorithm with BnB and LP (termed as BnB+LP). It is shown that the speed of the original algorithm is too slow for real time applications, especially when p is large. BnB and LP schemes significantly reduce the computation time. When $p = 6$, the BnB+LP implementation is 6.4 times faster than the original algorithm and this ratio tends to increase as p increases.

We next extensively test the speed performance of the implementation BnB+LP. Fig. 6 illustrates the relationship between the computation time and the number of frames p with different settings of ϵ. It is shown that the computation time increases in a quadratic manner as p increases, which is consistent with our analysis. Fig. 7 illustrates the relationship between the computation time and the number of requests n. It is shown that the computation time increases in a linear manner as n increases, which is also consistent with our analysis.

Fig. 8 shows how the output of the algorithm for a fixed set of inputs (n=10) changes when p increases from 1 to 4. It shows that our algorithm reasonably allocates the camera frames in each case. Consequently, the overall satisfaction increases as p increases.

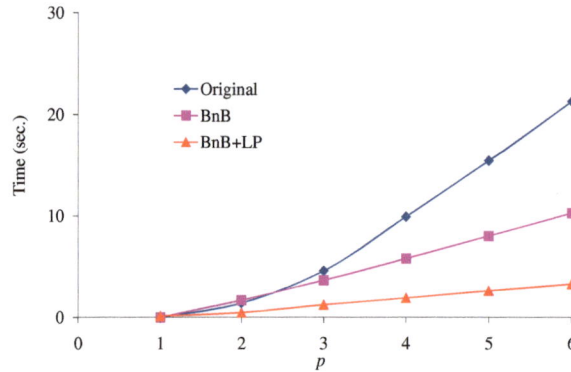

FIGURE 5. The computation time of different implementations vs. the number of frames p, (ϵ=0.25)

Evaluating System by Simulation. We carry out a simulation for evaluating the scheduling method of the system based on random inputs. The results are compared with an existing scheduling algorithm.

Simulation Setup. As shown in Fig. 9, a simulated $80 \times 60 \ m^2$ scene is constructed. In the first simulation setting, each object enters the scene through one side and maintains a constant speed. Seven random numbers are needed to characterize each object. First, a random integer number ranging from 1 to 4 is generated to indicate which side the object

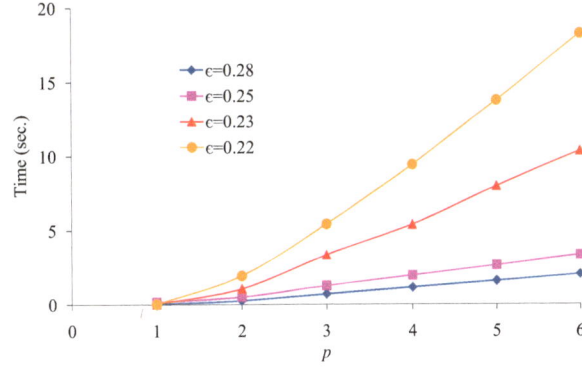

FIGURE 6. The computation time vs. the number of frames p, (n=50).

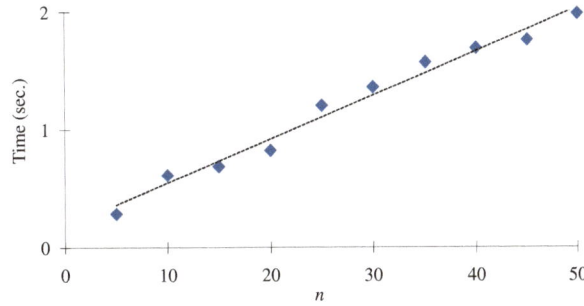

FIGURE 7. The computation time vs. the number of requests n, (p=4, ϵ=0.25).

enters through. Then a random real number in $[0, 1]$ is generated to indicate the entering point along the side. After that, the orientation of the object is determined by a random angle within the range $[-40°, 40°]$ with respect to the perpendicular of the side. The object speed is generated from a truncated Gaussian with a mean of 1.5 m/s and a standard deviation of 0.5 m/s, which is basically the speed of a walking people. The width and height of the rectangle that represents the object are randomly generated from a range $[1.5, 2.5]$ m. Finally, the desirable resolution of the object is generated from a range $[1, 21]$ (level), which is also the Panasonic HCM280 camera zoom range. The cameras run in 10 fps, which means $\tau = 0.1$ s. Then $\alpha = 30°$ and $\beta = 20°$. 5000 objects arrive in the scene following a Poisson process with arrival rate λ, which represents the congestion level of the scene. We set the lead time $\delta_l = 4$ s, which guarantees that in the request assignment phase, camera adjustment is completed before cameras intercept the objects. We set $\delta_r = 6$ s, which is equivalent to $n_r = 60$ frames. We set the parameter $n_e = n_r$ in (13) and $\rho = 0.5$ in (11) and $\epsilon = 0.25$. Two PTZ cameras are used, i.e., $p = 2$.

In the second simulation setting as shown in Fig., every κ time, the moving object changes its orientation by an angle uniformly in $(-\theta, \theta)$ with respect to its existing orientation. This generates an irregular, piece-wise linear trajectory, which is an approximation of the nonlinear trajectory of pedestrians in real world. In our simulation, we set $\kappa = \delta_l$ and $\theta = 5°$.

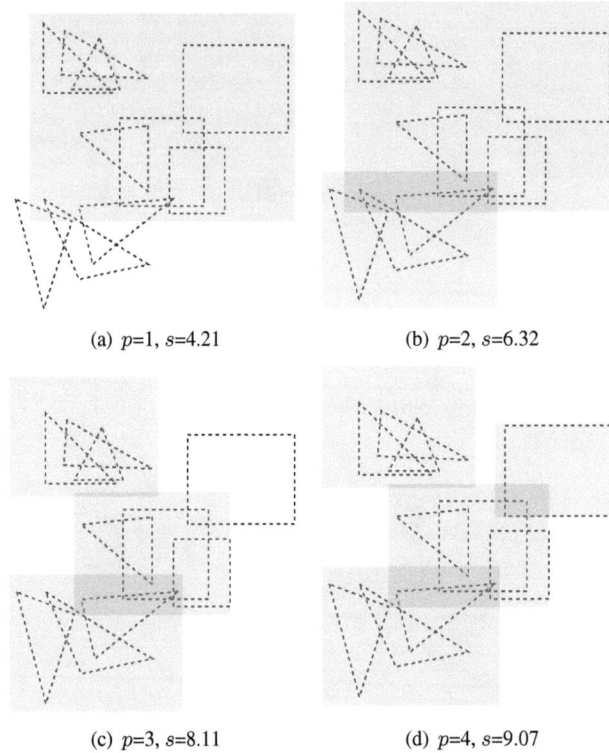

(a) $p=1$, $s=4.21$

(b) $p=2$, $s=6.32$

(c) $p=3$, $s=8.11$

(d) $p=4$, $s=9.07$

FIGURE 8. Sample outputs when p increases for a fixed input set $n = 10$.

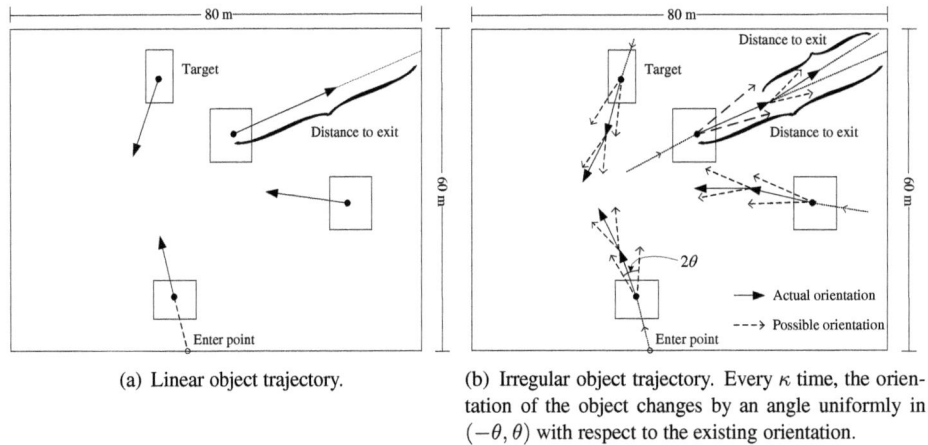

(a) Linear object trajectory.

(b) Irregular object trajectory. Every κ time, the orientation of the object changes by an angle uniformly in $(-\theta, \theta)$ with respect to the existing orientation.

FIGURE 9. An illustration of the simulated scene. Each object is represented as a rectangle and enters the scene from one of the four sides following a Poisson process. The initial orientation with respect to the norm of the side is bounded within $[-40°, 40°]$. The object maintains constant speed and its time to exit the scene is predicted.

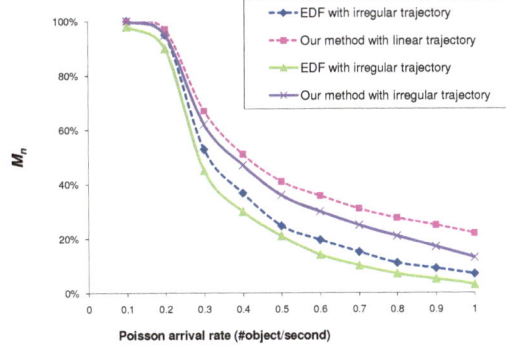

FIGURE 10. Comparison of scheduling policies based on M_n.

Metric and Results. We compare our scheduling scheme with the earliest deadline first (EDF) policy proposed in [6]. EDF is a heuristic scheme where the camera always picks the object with earliest deadline. With each congestion setting, 20 trials are carried out for average performance. We first compare the two schemes based on the ratio of number of objects that are observed for at least $n_r/2$ times to the total number of objects pass through the scene. We term this metric as M_n. This metric essentially indicates how many objects the system can capture and observe for a period of time. Fig. 10 shows the comparison result. It is shown that when the Poisson arrival rate λ is small, i.e., there are few objects in the scene, both scheduling schemes can reach almost best possible ratio (100%) under both trajectory settings. When λ increases, i.e., the traffic in the scene becomes heavy, the performance of EDF deteriorates significantly quicker than our method. It is also shown that the performance of both approaches under linear trajectory is better than that under irregular trajectory. This is due to the inaccuracy in object pose prediction caused by the irregularity. In the heavy traffic and irregular trajectory scenario, our method outperforms the EDF by over 200%.

We also compare our method with EDF based on the satisfaction to the objects since it takes into account not only the times that an object is observed, but also the resolution of the observation. As mentioned earlier, the AUS as defined in (12) indicates how well an object is satisfied. We define the second metric M_s as the ratio of average AUS to the maximum possible satisfaction for each object (i.e., n_e). Fig. 11 summarizes the comparison based on M_s. It is shown that our method outperforms EDF as λ increases under both trajectory settings. In the heavy traffic, irregular trajectory scenario, our method outperforms the EDF by about 300 %. This is not surprising since in heavy traffic situations, objects tends to be close to each other, where multi-object coverage has great advantage.

Careful analysis reveals that our satisfaction formulation in (4) is actually a generalization of many existing scheduling schemes. For example, if we tune parameter ρ in (11) to approach to zero, then the change in $\mu_i(t)$ dominates the change in the overall weight. That means we extremely care the emergency of the request and thus the scheduling converges to the earliest deadline first (EDF) policy [6]. Also, given we set the requested resolution close to the highest camera resolution, or we change the resolution ratio term $\min(\frac{z_i}{z}, 1)$ in (4) to indicator function $I(z_i \geq z)$. This means we only accept the images with the least requested resolution. Then the frame selection algorithm would assign at

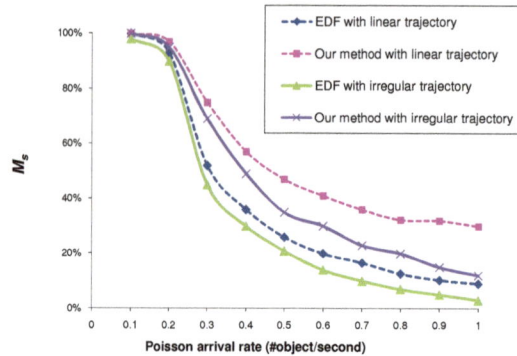

FIGURE 11. Comparison of scheduling policies based on M_s.

least a single object to each PTZ camera in the worst case, which is exactly the scheduling scheme based on the single object tracking as in almost all existing works.

Physical Experiment. We carry out a physical experiment to validate our system using real video data. Our camera is mounted on the 6th floor of the Evans Library of Texas A&M University to monitor the crowd entering and leaving the library. In the experiment, we set $t_0 = \delta_l = 1.5\ s$, $\delta_r = 2\ s$ and $p = 3$. The camera runs at 10 fps. Fig. 12 shows the key frames of a representee video clip that records an observation operation which contains two consecutive observation cycles at 17:25 on May 4th, 2009. It is shown that the request assignment module is capable of partitioning the objects and assigning each PTZ camera with a subset of the objects. The PTZ camera parameter selection module ensures the assigned objects are covered for the duration of the observation cycle. Between the observation cycles, the system also shows the ability to adjust the priority of the objects through the dynamic weighting so that every moving object is evenly observed.

7. Conclusion and Future Work

In this chapter, we present an autonomous vision system that consists of multiple robotic PTZ cameras and a fixed wide-angle camera for observing multiple objets simultaneously. We present the system with observation request generation, request assignment and PTZ camera parameter selection modules. We formulate the PTZ camera scheduling as a sequence of request assignment and camera parameter selection problems with objective of maximizing the satisfaction to requests. We propose an approximation algorithm to solve the problems. We test the speed of the algorithm in simulation. We validate the system by both simulation and physical experiments. The comparison with an existing work based on simulation shows that our system significantly enhances the observation performance especially in heavy traffic situations.

In the future, we will investigate how different frame selection formulation would impact the system performance and how they fit human user need in practice. Another interesting extension is to consider the camera traveling time within the request assignment. Intuitively, asynchronized observation by multiple PTZ cameras would further enhance the system performance. The camera content delivery through internet would be another interesting topic especially when number of camera increases.

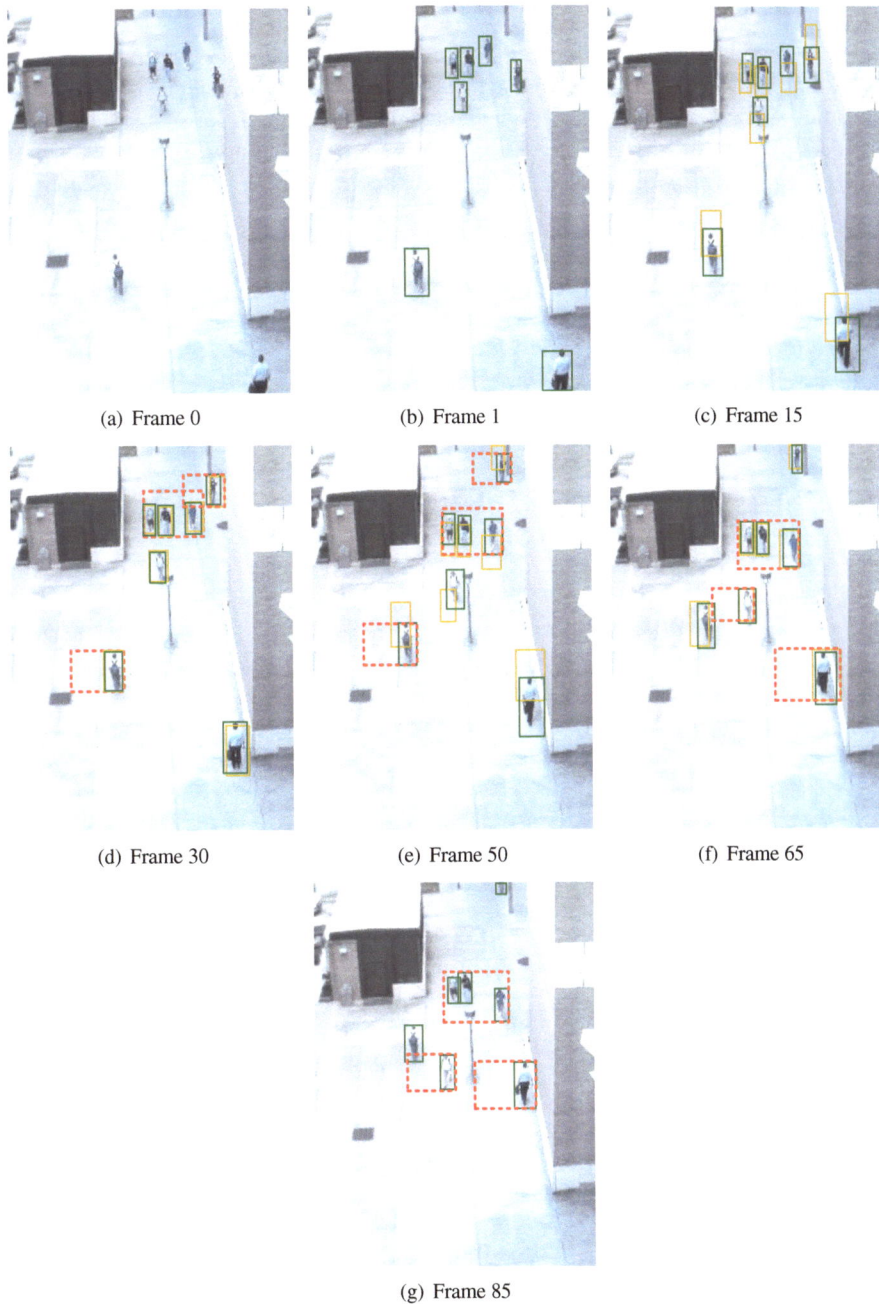

(a) Frame 0   (b) Frame 1   (c) Frame 15

(d) Frame 30   (e) Frame 50   (f) Frame 65

(g) Frame 85

FIGURE 12. Key frames in a representative surveillance cycle. (a) At time $t = 0$, there are 7 people in the initial scene. (b) The system starts to track the moving people, who are represented by green rectangles. (c) At time $t = t_0$, the states of the people at time $t = t_0 + \delta_l$ are predicted, which are represented by yellow rectangles. (d) At time $t = t_0 + \delta_l$, each PTZ camera is assigned a subset of the people. The optimal PTZ camera settings are represented by red dashed rectangles. (e) At time $t = t_0 + T$, one observation cycle finishes and the system predicts the states of the people at time $t = t_0 + T + t_l$ for the next cycle. (f) At time $t = t_0 + T + t_l$, each PTZ camera is again assigned a subset of the people. The better satisfied objects in the previous cycle are deprioritized through the dynamic weighting. (g) At time $t = t_0 + 2T$, the second observation cycle finishes.

Acknowledgement

We would like to thank N. Amato, J, Chen, F. van der Stappen, N. Papanikolopoulos, R. Volz and K. Goldberg for their insightful input, Q. Ni for implementing the motion detection, and C. Kim, J. Zhang, A. Aghamohammadi B. Qian, S. Hu and Z. Bing for their contribution to the Networked Robot Lab at Texas A&M University. We also acknowledge the funding support from National Science Foundation under IIS-0643298 and Microsoft.

Bibliography

[1] N. Krahnstoever, T. Yu, S.-N. Lim, K. Patwardhan, and P. Tu, "Collaborative real-time control of active cameras in large scale surveillance systems," in *Proc. Workshop on Multi-camera and Multi-modal Sensor Fusion Algorithms and Applications (M2SFA2)*, Marseille, France, October 2008.

[2] X. Zhou, R. T. Collins, T. Kanade, and P. Metes, "A master-slave system to acquire biometric imagery of humans at distance," in *IWVS '03: First ACM SIGMM international workshop on Video surveillance.* New York, NY, USA: ACM, 2003, pp. 113–120.

[3] A. Hampapur, S. Pankanti, A. Senior, Y.-L. Tian, L. Brown, and R. Bolle, "Face cataloger: multi-scale imaging for relating identity to location," *Proceedings. IEEE Conference on Advanced Video and Signal Based Surveillance, 2003.*, pp. 13–20, July 2003.

[4] R. Bodor, R. Morlok, and N. Papanikolopoulos, "Dual-camera system for multi-level activity recognition," *Intelligent Robots and Systems, 2004. (IROS 2004). Proceedings. 2004 IEEE/RSJ International Conference on*, vol. 1, pp. 643–648 vol.1, Sept.-2 Oct. 2004.

[5] L. Fiore, D. Fehr, R. Bodor, A. Drenner, G. Somasundaram, and N. Papanikolopoulos, "Multi-camera human activity monitoring," *J. Intell. Robotics Syst.*, vol. 52, no. 1, pp. 5–43, 2008.

[6] C. J. Costello, C. P. Diehl, A. Banerjee, and H. Fisher, "Scheduling an active camera to observe people," in *VSSN '04: Proceedings of the ACM 2nd international workshop on Video surveillance & sensor networks.* New York, NY, USA: ACM, 2004, pp. 39–45.

[7] S.-N. Lim, L. Davis, and A. Elgammal, "Scalable image-based multi-camera visual surveillance system," *Proceedings. IEEE Conference on Advanced Video and Signal Based Surveillance, 2003.*, pp. 205–212, July 2003.

[8] A. D. Bimbo and F. Pernici, "Distant targets identification as an on-line dynamic vehicle routing problem using an active-zooming camera," *Visual Surveillance and Performance Evaluation of Tracking and Surveillance*, pp. 97–104, 2005.

[9] F. Qureshi and D. Terzopoulos, "Smart camera networks in virtual reality," *Proceedings of the IEEE*, vol. 96, no. 10, pp. 1640–1656, Oct. 2008.

[10] S.-N. Lim, L. S. Davis, and A. Mittal, "Constructing task visibility intervals for video surveillance," *Multimedia Systems*, vol. 12, no. 3, pp. 211–226, 2006.

[11] S.-N. Lim, L. Davis, and A. Mittal, "Task scheduling in large camera networks," in *8th Asian Conference on Computer Vision*, Tokyo, Japan, 2007, pp. 397–407.

[12] C. Zhang, Z. Liu, Z. Zhang, and Q. Zhao, "Semantic saliency driven camera control for personal remote collaboration," *Multimedia Signal Processing, 2008 IEEE 10th Workshop on*, pp. 28–33, Oct. 2008.

[13] E. Sommerlade and I. Reid, "Information-theoretic active scene exploration," in *IEEE Conference on Computer Vision and Pattern Recognition (CVRP)*, Anchorage, Alaska, USA, 2008.

[14] N. Megiddo and K. Supowit, "On the complexity of some common geometric location problems," *SIAM Journal on Computing*, vol. 13, pp. 182–196, February 1984.

[15] D. Eppstein, "Fast construciton of planar two-centers," in *Proc. 8th ACM-SIAM Sympos. Discrete Algorithms*, January 1997, pp. 131–138.

[16] E. M. Arkin, G. Barequet, and J. S. B. Mitchell, "Algorithms for two-box covering," in *Symposium on Computational Geometry*, June 2006, pp. 459–467.

[17] H. Alt, E. M. Arkin, H. Brönnimann, J. Erickson, S. P. Fekete, C. Knauer, J. Lenchner, J. S. B. Mitchell, and K. Whittlesey, "Minimum-cost coverage of point sets by disks," in *Symposium on Computational Geometry*, June 2006, pp. 449–458.

[18] D. Song, *Sharing a Vision: Systems and Algorithms for Collaboratively-Teleoperated Robotic Cameras.* Springer, 2009.

[19] D. Song, A. F. van der Stappen, and K. Goldberg, "Exact algorithms for single frame selection on multi-axis satellites," *IEEE Transactions on Automation Science and Engineering*, vol. 3, no. 1, pp. 16–28, January 2006.

[20] D. Song and K. Goldberg, "Approximate algorithms for a collaboratively controlled robotic camera," *IEEE Transactions on Robotics*, vol. 23, no. 5, pp. 1061–1070, October 2007.

[21] D. Song, N. Qin, and K. Goldberg, "Systems, control models, and codec for collaborative observation of remote environments with an autonomous networked robotic camera," *Autonomous Robots*, vol. 24, no. 4, pp. 435–449, 2008.

[22] A. Elgammal, D. Harwood, and L. Davis, "Non-parametric model for background subtraction," in *6th European Conference on Computer Vision (ECCV)*, vol. 2, Dublin, Ireland, June/July 2000, pp. 751–761.

[23] A. Yilmaz, O. Javed, and M. Shah, "Object tracking: A survey," *ACM Computing Surveys*, vol. 38, no. 4, pp. 1–45, 2006.

Send Orders of Reprints at reprints@benthamscience.net
Networking Humans, Robots and Environments, 2013, 43-58   **43**

CHAPTER 3

Distributed Sensing and Human-Aware Robot Reasoning Mechanisms

Fulvio Mastrogiovanni, Antonio Sgorbissa, and Renato Zaccaria

Department of Computer, Communication and System Sciences

University of Genova

Genova, Italy

ABSTRACT. This Chapter introduces a formal language used to model complex relationships between detectable human activities, events and robot behaviours that are to be detected by robots in streams of sensory data. The Ubiquitous Robotics paradigm is adopted in order to show how distributed sources of information can be easily integrated to assess the current context. Specific emphasis is devoted to the use of context assessment techniques to enforce the adaptation of robot behaviours to human activity. The model is described both theoretically and with a thorough example.

1. Introduction

According to the *Ubiquitous Robotics* paradigm [**5, 11**], mobile robots are part of a fully *networked* system that is *enriched* by intelligent devices (both sensors and actuators) distributed throughout the environment. Networked Robots can therefore cooperate with intelligent devices to perform tasks requiring sophisticate cognitive or physical interaction capabilities, such as *distributed sensing* (e.g., integrating on board and remote sensory information) or skilled actuation (e.g., simply requesting automated doors to open instead of using fine manipulation capabilities to operate on door handles). In these cases, highly specialized intelligent devices can provide with information exchange for the lack of cognitive and physical interaction capabilities traditionally exhibited by robots [**13**]: "autonomy" and "situatedness" refer to the whole system, rather than on the individual robotic platform. It is widely recognized that this integration is expected to improve capabilities, reliability and performance of real-world robotic systems, and in particular to enforce (possibly implicit) Human-Robot interaction processes [**14**].

Architectures proposed in the literature lack in addressing how this novel paradigm affects context assessment processes, specifically in the case where human activity must be detected, monitored and "understood", with the aim of tuning robot reasoning accordingly. When aggregating heterogeneous information originating from distributed sources, the need arises for a theoretical framework investigating the following issues:

- Context assessment strategies to deal with sensory data collected from intelligent devices as well as teams of mobile robots.
- A principled study of the impact arising from the integration between *embodied* knowledge representation (which is robot-dependent and mobile) and distributed knowledge representation (which is device-dependent and localized in space).

A promising approach to deal with distributed context assessment is to describe contexts using high-level symbolic frameworks, which are responsible for managing both the acquisition and the interpretation of patterns of sensory data. This Chapter describes ongoing work dealing with such issues as distributed knowledge representation, context assessment and human activity recognition in Ubiquitous Robotics scenarios, where mobile robots and intelligent environments cooperate to build models for human activity and to detect relevant patterns of occurring events, possibly generated by human behaviour (see Figure 1). These models are based on symbolic structures that mutually maintain semantic relationships with respect to each other and are grounded by sensory data as soon as they are available. The main contribution of the Chapter is the description of the Situation Definition Language (henceforth referred to as $\mathcal{SDL}$), which is aimed at easing the process of specifying and representing relevant symbolic structures, such as situations to recognize, as *sequences of events and detectable human activities*. Stemming from previous work [**12**],

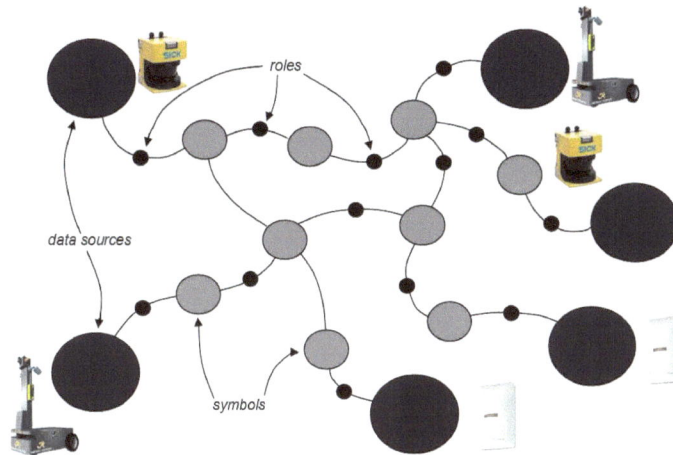

FIGURE 1. A sketch of the distributed architecture for context assessment.

the language adopts a model for context awareness integrating *ontology* and logic-based approaches.

The Chapter is organized as follows. Section 2 briefly discusses situation languages for context models. Section 3 describes $\mathcal{SDL}$ both at formal and semantic levels. Section 4 introduces the context assessment procedure. Finally, Section 5 describes a significant example. Conclusions follow.

2. Related Work

The notion of "context" [3, 4, 7] has been defined as "any information that can be used to characterize the situation of an entity". Models for endowing Ubiquitous Robotics scenarios with context awareness are subject to three requirements [6, 16, 18]:

- *effectiveness* in assessing sensory information;
- *reusability* in aggregating data to generate composite structures;
- *expressiveness* in representing human activities and other events taking *temporal* relationships into account.

Such modelling tools as *ontologies* are considered a good candidate solution in many real-world applications. However, it is necessary to constrain their modelling capabilities (e.g., defining symbols with special meaning) in order to describe sequences of events. This process originates a *language* able to encode relationships between different ontology symbols representing templates of events that must be searched for when processing sensory data at run time. Among these, four have been identified for this discussion.

The well-known "Context Toolkit" [3] is based on three language elements, namely context *widget*, *interpreter* and *aggregator*. Widgets are associated with real-world artefacts devoted to acquire sensory data and to make it accessible for further processing through services. Interpreters correspond to modules reasoning upon contextual knowledge (i.e., sensory data as well as internal data structures): they can process heterogeneous information and produce contextual knowledge. Finally, aggregators are modules aimed at combining heterogeneous contextual knowledge.

The "Context Broker Architecture" (CoBrA in short) and the associated ontology [1] originate from well-known *desiderata* of distributed computing systems: on one hand, contextual knowledge must be made available to the whole network; on the other hand, it must be provided with a well-defined and shared meaning. These two aspects are realized by a *broker* in charge of:

- maintaining a shared language used to model situations;
- acquiring sensory data from distributed information sources;
- establishing information delivery policies to share contextual knowledge.

A general purpose context model, referred to as the "Aspect-Scale-Context" model, along with an associated situation language, namely the Context Ontology Language (i.e., CoOL) [15] originates from the *shared understanding* issue characterizing distributed systems: each participant "actor" must be provided with the same semantics about the data represented globally within the system. Therefore, CoOL is structured to enforce interoperability: situations are organized as a set of particular "aspects", each one represented with respect to one or more "scales", which relate different facets of represented contextual knowledge. However, the model has two major drawbacks: on one hand, a pairwise mapping between different scales must be available; on the other hand, a metric must be provided to establish the mapping.

A hierarchical situation language, enforcing temporal relationships between symbols, is discussed in [8]. In this framework, a context is a sequence of "context states" depending on "context features", which are individual and atomic language elements. Context features must be carefully assessed and evaluated to determine which features are relevant for a particular context to model.

Although many approaches for implementing Ubiquitous Robotics scenarios have been presented in literature [11], few of them explicitly introduce models for implementing context awareness capabilities. Within the *Ambience* project [17], mobile robots are coupled with smart environments to manage tasks related to human-robot interaction and emotional awareness. However, a principled context language is not considered for defining Human-Robot interaction processes, thereby lacking in reusability and expressiveness. The *Ubibot* paradigm [5] is aimed at realizing ubiquitous systems composed of three main constituents: *Mobots* (i.e., mobile robots), *Embots* (i.e., embedded robots: agents in charge of performing data acquisition) and *Sobots* (i.e., software robots: agents implementing computational and cognitive algorithms). A properly defined context language requires a tight integration at the behavioural level between Mobots, Embots and Sobots. However, contextual knowledge is used only to maintain a very simple model of *user* contexts, which lack in effectiveness and reusability. Finally, as part of the *PEIS Ecology* framework, and grounded with respect to the specific field of *cooperative anchoring*, the work described in [2] is one of the first attempts to generate composite conceptual structures originating from distributed sensory data, which explicitly take into account temporal ordering between events. Unfortunately, issues related to *reusability* and *expressiveness* are not sufficiently structured for the purpose of properly defining a context language.

3. A Language to Describe Sequences of Events

This Section introduces the $\mathcal{SDL}$ language, which allows to define sequences of events (including those generated by human activity) that, once recognized by a context aware system [12], can be used to trigger specific system behaviours. $\mathcal{SDL}$ is obtained by constricting the representation capabilities of an ontology Σ (i.e., a set of concepts $\{\sigma_k\}$ and the relationships between those concepts, each one provided with a description $\mathcal{D}_k$), and it benefits from associated inference mechanisms, for instance *subsumption* (i.e., a "set-inclusion" like operator acting upon two descriptions), which is used to classify patterns of sensory data according to the defined sequences.

3.1. Definitions. Symbols in the set $\{\sigma_k\}$, $k = 1, ..., |\sigma_k|$, are divided into four classes:

- *Basic Types* $\mathcal{B}$ representing sensory data;
- *Predicates* $\mathcal{P}$ asserting facts using basic types: they are the lowest layer of the knowledge representation system since they interface it with *information sources* (see Figure 1), which can be sensors (and associated processing algorithms) either distributed in the environment or on board the robots;
- *Contexts* $\mathcal{C}$ meant at composing *Predicates* to relate the associated information: *Contexts* arrange *Predicates* and simpler *Contexts* referring to the same subject of attention (e.g., a user, a robot or an element in the environment) to relate heterogeneous sensory data;
- *Situations* $\mathcal{S}$ considering aggregates of *Contexts*: they model sequences of symbols representing sensory data templates that must be recognized at run time.

Within the ontology, symbols σ_k are associated with descriptions $\mathcal{D}_k$, which are always expressed in *conjunctive normal form*, i.e., in infix notation,

$$\mathcal{D}_k \leftarrow \mathcal{D}_{k,1} \sqcap ... \sqcap \mathcal{D}_{k,i} \sqcap ... \sqcap \mathcal{D}_{k,n}.$$

A description $\mathcal{D}_k$ corresponds either to a $\mathcal{B}$, $\mathcal{P}$, $\mathcal{C}$ or $\mathcal{S}$ symbol, or it can be iteratively composed on the basis of other descriptions to build higher level *Predicates*, *Contexts* or *Situations*.

The process of encoding sensory information is managed through a *variable assignment* α, which substitutes actual sensory values to the corresponding basic types variables in a description $\mathcal{D}_k$. A grounded description $\hat{\mathcal{D}}_l$ originates from α according to the definition of $\mathcal{D}_k$, and hence corresponds to an instance of a basic type, or it can be iteratively composed using other grounded descriptions to build instances of *Predicates*, *Contexts* or *Situations*. On one hand, given an interpretation $\mathcal{I}$ and a variable assignment α over basic types such that

$$(\mathcal{P}, \mathcal{C}, \mathcal{S}, \mathcal{I}, \alpha) \models \mathcal{D}_k(\hat{\mathcal{D}}_l),$$

the description $\mathcal{D}_k$ can be satisfied for $\mathcal{I}$ and α; on the other hand, grounded symbols $\hat{\sigma}_l$ originate from a variable assignment α according to the definition of a given $\mathcal{D}_k$, i.e., $\hat{\mathcal{D}}_l \models \sigma_k(\hat{\sigma}_l)$.

Since each *grounded symbol* ultimately represents detected human activities or events, a computational process extracting simple contexts from sensory data and assessing semantically important information from combinations of simple contexts is implicitly defined, once an interpretation of the symbol is provided.

Adopting the same mechanism, it is possible to take *temporal* relationships into account, by assuming that time is modelled as a discrete ordered sequence of instants τ_i. This allows us to redefine α_τ as the process of grounding basic type variables anchored to sensory data at the time instant τ in a description $\mathcal{D}_k$. Two formal implications can be pointed out: on one hand, given a time instant τ, an interpretation $\mathcal{I}$ and a variable assignment α_τ over basic types such that

$$(\mathcal{P}, \mathcal{C}, \mathcal{S}, \mathcal{I}, \alpha_\tau) \models \mathcal{D}_k(\hat{\mathcal{D}}_{l,\tau}),$$

the description $\mathcal{D}_k$ can be satisfied in τ for $\mathcal{I}$ and α_τ; on the other hand, given a time instant τ, grounded symbols $\hat{\sigma}_{l,\tau}$ originate from a variable assignment α_τ according to the definition of $\mathcal{D}_k$, i.e., $\hat{\mathcal{D}}_{l,\tau} \models \sigma_k(\hat{\sigma}_{l,\tau})$.

The *subsumption* operator is the core of the inference process used to assess situations at run time. This classification process matches streams of sensory data with respect to $\mathcal{SDL}$ formulas: if this stream is subsumed by at least one $\mathcal{SDL}$ formula, the formula is satisfied and the associated encoded pattern of human activities or events is recognized. Subsumption is a binary operator (henceforth referred to as subs?) that, given two descriptions $\mathcal{D}_1$ and $\mathcal{D}_2$, returns *true* whenever $\mathcal{D}_1$ is *more general* or *equivalent* to $\mathcal{D}_2$:

$$\{true|false\} \leftarrow \text{subs?}(\mathcal{D}_1, \mathcal{D}_2).$$

Input terms can be either non-grounded or grounded descriptions. In other words, the expression subs?$(\mathcal{D}_1, \mathcal{D}_2)$ returns *true* if and only if, for each possible interpretation $\mathcal{I}$ and variable assignment α it holds that $\mathcal{D}_1$ is satisfied whenever $\mathcal{D}_2$ is satisfied. Given a time instant τ, an interpretation $\mathcal{I}$ and a variable assignment α_τ, and given a set of *Predicates* $\{\hat{\sigma}_{l,\tau}\}$ grounded with respect to symbols $\{\sigma_k\}$ at instant τ, and their corresponding descriptions $\{\hat{\mathcal{D}}_{l,\tau}\}$, then *situation assessment* requires to perform queries in the form subs?$(\mathcal{D}_k, \hat{\mathcal{D}}_{l,\tau})$ for each relevant $\mathcal{D}_k$, i.e., the descriptions associated with *Contexts* and *Situations* that must be checked to hold at τ.

$\mathcal{SDL}$ can be defined as

$$\mathcal{SDL} = L(\mathcal{SDL}_{st}),$$

where L is a generative process operating on a grammar $\mathcal{SDL}_{st}$. $\mathcal{SDL}_{st}$ is formally defined as a 2-ple

$$\mathcal{SDL}_{st} = <A, R>,$$

where A is a finite *alphabet* and R is a binary relationship over A^* (the set of finite, possibly empty, strings over A) such that $R \subseteq A^* \times A^*$. The grammar specifies how to build well-formed *sentences* belonging to $\mathcal{SDL}$ starting from an initial sentence in A^* and iteratively applying *rewrite rules* belonging to R until the intended sentence is *derived* in n steps. Sentences s_i in $\mathcal{SDL}$ are elements of the set A^*, whereas rewrite rules $r_a \to r_c$ are defined as binary elements $(r_a, r_c) \in R$, where r_a and $r_c \in A^*$ are called, respectively, the *antecedent* and the *consequent* of the rule.

The generative process L must be precisely defined. Preliminarily, given a_1 and $a_2 \in A^*$, a one-step rewrite relation $\Rightarrow_R$ over an alphabet A^* can be defined such that, given x, y, r_a and $r_c \in A^*$, then $a_1 \Rightarrow_R a_2$ holds if and only if $a_1 = x r_a y$, $a_2 = x r_c y$ and $(r_a, r_c) \in R$. Then, an n-derivation in $\mathcal{SDL}$ is defined as a finite sequence of sentences $s_0, ..., s_n$ that are produced by starting from an initial sentence s_0 belonging to A^* and rewriting it by means of rewrite rules in R. As a consequence, the generative process L can be defined such that

$$\mathcal{SDL} = \{s_i \in A^* | s_0 \Rightarrow_R s_i\},$$

from which it can be deduced that $\mathcal{SDL}$ is a Semi-Thue system.

Sentences are not atomic elements of $\mathcal{SDL}$. On the contrary, they can be decomposed into many parts, which are described in the following paragraphs. To this purpose, the alphabet A is defined as a set of symbol classes such that

$$A = \{s_0, s, \emptyset, l, f, \oplus, t, v, c\}.$$

It is important to point out that symbols in A represent different classes of language elements in $\mathcal{SDL}$. In particular:

- s_0, s and $\emptyset$ are, respectively, the initial sentence, the generic sentence and the null sentence;
- l represents *labels*, meant at providing each sentence with a name;
- f represents *formulas*, i.e., building blocks upon which sentences can be built;
- $\oplus$ represents both unary and binary connecting *operators*;
- t represents *terms*, i.e., everything else apart from sentences, and more precisely language *variables* v and *constants* c.

A *sentence* s is an element of A^* in the form $l \equiv l(f);$, where l represents suitable labels and f is a formula. A *formula* f is an element of A^* that can be a label, a term, a juxtaposition or a composition of formulas by means of connecting operators. Finally, a *connecting symbol* $\oplus$ is an element of A that contributes to formulas by composing one or more subformulas using, respectively, *unary* symbols $\oplus_u$ and *binary* symbols $\oplus_b$.

It is possible to further specify the set of rewrite rules R that characterizes the $\mathcal{SDL}$ language. In particular, these are defined as follows:

- $s_0 \to s \mid ss_0 \mid \emptyset;$
- $s \to l \equiv l(f);;$
- $f \to l \mid t \mid \oplus_u f \mid f \oplus_b f;$
- $l \to \texttt{string};$
- $t \to v \mid c;$
- $v \to \texttt{string};$
- $c \to \texttt{string};$
- $\oplus_u \to \neg \mid \equiv \mid \delta \mid ();$
- $\oplus_b \to \sqcap \mid \sqcup \mid \lambda_\omega \mid \prec \mid.$

The syntax of $\mathcal{SDL}$ is completely specified on the basis of the alphabet A and the rewrite rules in R. However, what is unspecified is how to ground the language with respect to the underlying ontology Σ, and in particular to descriptions $\{\mathcal{D}_k\}$: it is necessary to map language elements to symbol classes within the ontology.

The rule $v \to \texttt{string}$ maps a set of variables v_k, $k = 1, ..., |v_k|$, to corresponding symbols within Σ related to *Basic Types* $\mathcal{B}$ such that $\{\sigma_k | \Sigma \models \mathcal{B}(\mathcal{D}_k)\}$. Variables v_k correspond to descriptions $\mathcal{D}_k$ referring to basic types. Furthermore, with $\texttt{string}$ in this case the name of the related σ_k is referred to. The rewrite rule $c \to \texttt{string}$ maps a set of constants c_l, $l = 1, ..., |c_l|$, to symbols within the ontology related to instances of *Basic Types* $\mathcal{B}$ such that $\{\hat{\sigma}_l | \Sigma \models \mathcal{B}(\hat{\mathcal{D}}_l)\}$. Differently from variables, constants c_l correspond to descriptions $\hat{\mathcal{D}}_l$ that refer to instances of basic types. However, $\texttt{string}$ refers to the name of the related $\hat{\sigma}_l$. On the other hand, the rule $l \to \texttt{string}$ maps labels to a set of named non-grounded σ_k or grounded $\hat{\sigma}_l$ symbols within the ontology Σ, each one related to either *Predicates*, *Contexts* or *Situations*. These symbols contribute to the definition of formulas. Specifically, the previously defined rewrite rule $f \to l \mid t \mid \oplus_u f \mid f \oplus_b f$ maps formulas f_j, $j = 1, ..., |f_j|$ to corresponding either non-grounded or grounded descriptions within Σ, which are related to either *Predicates*, *Contexts* or *Situations*.

Predicates originate from the previously introduced rewrite rule $s \to l \equiv l(f);$, which maps a set of sentences $p_j \in A^*$, $j = 1, ..., |p_j|$, to corresponding symbols within Σ. These symbols are related to *Predicates* $\mathcal{P}$ which descriptions are such that $\{\sigma_k | \Sigma \models P(D_k)\} \cup \{\hat{\sigma}_l | \Sigma \models P(\hat{D}_l)\}$. Sentences p_j correspond to descriptions related to basic types in conjunctive normal form that describe few "facts" about an entity. Analogously, the rule $s \to l \equiv l(f);$ maps a set of sentences $c_j \in A^*$, $j = 1, ..., |c_j|$, to corresponding symbols within Σ related to *Contexts* $\mathcal{C}$ which descriptions are such that $\{\sigma_k | \Sigma \models \mathcal{C}(\mathcal{D}_k)\} \cup \{\hat{\sigma}_l | \Sigma \models \mathcal{C}(\hat{\mathcal{D}}_l)\}$. In this case, sentences c_j correspond to descriptions related to predicates in conjunctive normal form that refer to the same entity. Finally, also *Situations* are mapped: the rule $s \to l \equiv l(f)$; maps a set of sentences $s_j \in A^*$, $j = 1, ..., |s_j|$, to corresponding symbols within Σ related to *Situations* $\mathcal{S}$ which descriptions are such that $\{\sigma_k | \Sigma \models \mathcal{S}(\mathcal{D}_k)\} \cup \{\hat{\sigma}_l | \Sigma \models \mathcal{S}(\hat{\mathcal{D}}_l)\}$. Obviously enough, sentences s_j correspond to descriptions related to predicates in normal conjunctive form that refer to different entities.

3.2. Semantics. Every time instant τ, grounded formulas originate from a variable assignment α_τ according to a specific interpretation $\mathcal{I}$. Sentences in $\mathcal{SDL}$ assert which symbols must be grounded in τ for the represented sentence to be satisfied, given sensory information acquired from some instant in the past up to the present time. How the grounding process is performed depends on the constituent formulas. Ultimately, this requires to identify a proper interpretation $I \doteq (\Sigma, \cdot^I)$ for symbols belonging to the underlying ontology Σ, thereby imposing a semantics to the corresponding *formulas*.

In the following paragraphs, it is demonstrated first that, given a generic formula produced by rules in R, it is possible to build the corresponding parsing tree. This result is then used to determine how the variable assignment α_τ propagates throughout the formula itself. Finally, the meaning of each connecting symbol with respect to constituent formulas is described in detail, which is of the uttermost importance since actual formulas are encoded in an ontology.

Theorem. Given a finite-length sequence of either non-grounded or grounded symbols belonging to A^* (and built using rewrite rules as described the previous Section) it is possible to uniquely determine if such a sequence is a formula, to determine its logic sign, and to identify formulas.

Proof. For each formula f, it is possible to associate a finite-length tree, which is able to visually represent the process of formula derivation by rules in R until reaching a `string` element. The tree is generated using the following rules:

- the tree root is a node that is labelled with the given formula f;
- if a node is labelled by a *label*, then it is a leaf;
- if a node is labelled by a *term*, then it has one child node, which is labelled either as a variable or a constant;
- if a node is labelled by a *variable*, then it is a leaf;
- if a node is labelled by a *constant*, then it is a leaf;
- if a node is labelled by a formula in the form $\oplus_u f$, then it has one child node that is labelled using f;
- if a node is labelled by a formula in the form $f_1 \oplus_b f_2$, then it has two child nodes, which are labelled, respectively, using f_1 and f_2.

Simply parsing the tree demonstrates the theorem.

In virtue of the previously introduced Theorem, it is possible to draw a *top-down* direct parallelism between finite-length trees and the formulas introduced in the previous Section. Since each sentence in $\mathcal{SDL}$ represents a *Situation*, building the corresponding finite-length tree for the *constituent formulas* is equivalent to decompose it into constituent descriptions. However, once $\mathcal{SDL}$ formulas are mapped to the ontology, they can not be mapped any more to finite-length trees, since the same symbol can correspond – in principle – to different leaves. However, formulas encoded within an ontology can be still treated as Directed Acyclic Graphs.

In order to define an appropriate semantics, the effects of the variable assignment α over $\mathcal{SDL}$ formulas must be defined. The *substitution* of an actual basic type value α_i to all the occurrences of a variable v in a term t, that is referred to as $t[v/\alpha_i]$, is recursively defined as:

- if t is a variable s.t. $v \neq t$, then $t[v/\alpha_i] \rightarrow t$;
- if t is a variable s.t. $v = t$, then $t[v/\alpha_i] \rightarrow \alpha_i$;
- if t is a constant, then $t[v/\alpha_i] \rightarrow t$.

The substitution of a term t to all the occurrences of a variable v in a formula f, that is referred to as $f[v/t]$, is recursively defined as the following specification:

- if $f_{\oplus_u}$ is a formula in the form $\oplus_u f$, then $f_{\oplus_b}[v/t] \rightarrow \oplus_u f[v/t]$;
- if $f_{\oplus_b}$ is a formula in the form $f_1 \oplus_b f_2$, then $f_{\oplus_b}[v/t] \rightarrow f_1[v/t] \oplus_b f_2[v/t]$.

As soon as new information is available at the time instant τ either from distributed devices or robots, a new variable assignment α_τ is defined. Accordingly, variables are updated. As a consequence, corresponding formulas are given proper truth values, thereby satisfying more complex sentences.

Furthermore, the meaning associated with connecting operators must be clearly assessed. In order to recursively compose formulas, unary and binary symbols can be better classified in *relational* and *temporal* operators, that are characterized by an intuitive correspondence with common logic and mathematical operators.

Specifically, common logic operators have been added. The unary operator $\neg \in \oplus_u$ produces a formula $f_\neg \leftarrow \neg f$ that is the negation of the formula f. Given an interpretation $\mathcal{I}$ and a variable assignment α, $\neg \hat{f}$ is satisfied if and only if the logical expression *not* $\hat{f}$ is satisfied. An assignment operator is necessary to assign labels. In particular, the unary operator $\equiv \in \oplus_u$ produces a formula $f_\equiv \leftarrow \equiv f$ that is the copy of the formula f. Given an interpretation $\mathcal{I}$ and a variable assignment α, $\equiv \hat{f}$ is satisfied if and only if the logical expression $\hat{f}$ is satisfied. On the other hand, the unary operator $() \in \oplus_u$ produces a formula $f_{()} \leftarrow (f)$ that specifies the level of precedence in the parsing process for the formula f. Given an interpretation $\mathcal{I}$ and a variable assignment α, $(\hat{f})$ is satisfied if and only if the logical expression $\hat{f}$ is satisfied. Conjunction will be used to build complex formulas: the binary operators $\{\sqcap, , \} \in \oplus_b$ produce formulas $f_{\{\sqcap,,\}} \leftarrow f_1\{\sqcap,,\}f_2$ that are the juxtaposition of two formulas f_1 and f_2. Given an interpretation $\mathcal{I}$ and a variable assignment α, $\hat{f}_1\{\sqcap,,\}\hat{f}_2$ are satisfied if and only if the logical expression $\hat{f}_1$ *and* $\hat{f}_2$ is satisfied. It is worth noting that the correspondence of meaning between the two latter connecting operators originates from the mapping process of formulas to structures within the ontology. However, this is permitted to guarantee compatibility with commonly used logic formalisms. Finally, the binary operator $\sqcup \in \oplus_b$ produces a formula $f_\sqcup \leftarrow f_1 \sqcup f_2$ that is the disjunction of two formulas f_1 and f_2. Given an interpretation $\mathcal{I}$ and a variable assignment α, $\hat{f}_1 \sqcup \hat{f}_2$ is satisfied if and only if the logical expression $\hat{f}_1$ and $\hat{f}_2$ is satisfied.

Temporal operators require to take several time instants τ into account. As a consequence, they contribute to formulas that are satisfied by a sequence of several variable assignments α_τ. Three operators have been defined: *derivative*, *time length* and *precedence*. With respect to *derivative*, the unary operator $\delta \in \oplus_u$ produces a formula $f_\delta \leftarrow \delta(f)$ that is the "derivative" of the formula f. Given two time instants τ and $\tau - 1$, an interpretation $\mathcal{I}$ and two assignments α_τ and $\alpha_{\tau-1}$, $\delta_\tau(\hat{f}_\tau)$ is satisfied if and only if $\hat{f}_\tau = \neg \hat{f}_{\tau-1}$. For simplicity, a couple of operators, namely *positive derivative* $\delta^{t \rightarrow f}$ and *negative derivative* $\delta^{f \rightarrow t}$ are used in place of the general operator δ, in order to differentiate the case in which the argument of the derivative changes from *satisfied* to *not satisfied* or *vice-versa*. Next, the family of binary operators $\lambda_\omega \in \oplus_b$, $\omega \in \{<, \leq, =, \geq, >\}$, produces a formula

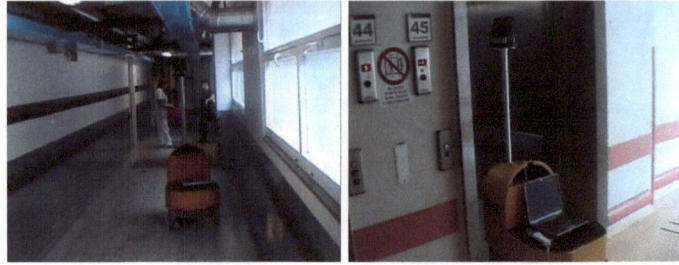

FIGURE 2. Mobile robots operating at the Polyclinic of Modena.

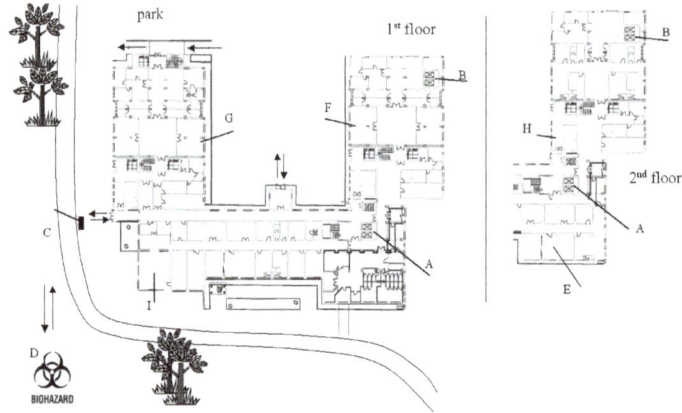

FIGURE 3. The operating environment: a) the ground floor and the park; b) the second floor.

$f_\lambda \leftarrow \lambda_\omega(f_1, f_2)$ that compares the temporal duration of two formulas f_1 and f_2. Given three time instants τ, τ_1 and τ_2, an interpretation $\mathcal{I}$ and three variable assignments α, α_{τ_1} and α_{τ_2}; given that $\tau_1 < \tau$ is the most recent time instant such that $\delta(f_1)$ holds, and $\tau_2 < \tau$ is the most recent time instant such that $\delta(f_2)$ holds; then the expression $\lambda_{\omega,\tau}(\hat{f}_{1,\tau}, \hat{f}_{2,\tau})$ is satisfied if and only if $(\tau - \tau_1)\omega(\tau - \tau_2)$ is satisfied, whereas it is not satisfied otherwise. In most cases, one of the constituent formulas may simply correspond to actual time intervals (e.g., "20 minutes" or "8 hours"), and the operator is currently used to check whether a formula lasts less, equally, or more than the specified interval. Finally, the binary operator $\prec \in \oplus_b$ produces a formula $f_\prec \leftarrow \prec (f_1, f_2)$ that expresses the temporal precedence relationship between two formulas f_1 and f_2. Given two instants τ_1 and τ_2, an interpretation $\mathcal{I}$ and two variable assignments α_{τ_1} and α_{τ_2}; given that $\delta^{t \rightarrow f}(\hat{f}_{1,\tau_1})$ and $\delta^{f \rightarrow t}(\hat{f}_{2,\tau_2})$ hold; then the expression $\hat{f}_{1,\tau_1} \prec \hat{f}_{2,\tau_2}$ is satisfied if and only if $\tau_1 < \tau_2$ holds (i.e., if the interval in which f_1 is satisfied strictly precedes the interval in which f_2 is satisfied), and not satisfied otherwise.

4. The Context Assessment Process

Context awareness is realized by aggregating *Predicate* instances in order to satisfy symbolic structures representing *Contexts* and *Situations* maintained within Σ. The aggregation assumes the form of the *history* of the n most recent *Predicate* instances P_i,

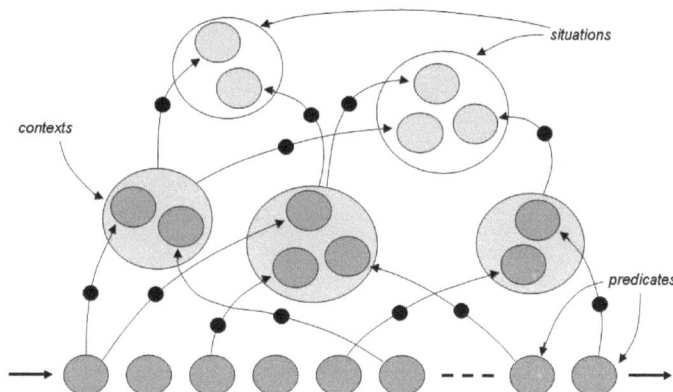

FIGURE 4.   The subsumption-based context assessment process.

$i = 1, ..., n$, which is stored within Σ in a *first-in-first-out* approach (see Figure 4 on the bottom).

At each time instant τ, when either new or updated sensory data are available, a classification process is carried out over *Predicate* instances, thus – possibly – modifying their truth value. As a matter of fact, *Predicate* instances can be either true of false according to specific sensory values. For instance, a *Crowded Predicate* can be satisfied by the continuous activation of a motion detection system on board of a camera device pointing in a given area. In particular, the overall history description $\mathcal{D}_p$ is considered; $\mathcal{D}_p$ is obtained by joining the description of each *Predicate* instance P_i:

$$\mathcal{D}_p \doteq \mathcal{D}(P_1 \sqcap ... \sqcap P_n).$$

$\mathcal{D}_p$ represents the system state with respect to both the current situation and the most recent past events. For the *Theorem* in Section 3.2, the system can infer what *Context* symbols C_j are satisfied. This is accomplished by checking $\texttt{subs?}[C_j, \mathcal{D}_p]$ between each *Context* and *history*. Specifically, $\mathcal{C}_s(\tau)$ is the collection of *Context* instances C_j subsuming $\mathcal{D}_p$:

$$\mathcal{C}_s(\tau) = \{C_j : \mathcal{D}_p \sqsubseteq C_j\}.$$

Therefore, all the satisfied $\texttt{Context}$ symbols $C_j \subseteq \mathcal{C}_s(\tau)$ are occurring in τ (see Figure 4 on the mid). This mechanism is easily iterated for *Situations*. Since $\mathcal{C}_s(\tau)$ varies at each τ, the history description $\mathcal{D}_c$ is considered, obtained by superimposing the description of each *Context* instance $C_j \subseteq \mathcal{C}_s(\tau)$:

$$\mathcal{D}_c \doteq \mathcal{D}(C_1 \sqcap ... \sqcap C_{|\mathcal{C}_s(\tau)|}).$$

Analogously to *Context* symbols, using $\mathcal{D}_c$ the system can infer what *Situation* elements S_k are satisfied in τ (see Figure 4 on top). Again, this is managed by $\texttt{subs?}[S_k, \mathcal{D}_c]$ for each S_k. Current satisfied *Situations* are stored in $\mathcal{S}_s(\tau)$, which is the collection of *Situation* instances S_k subsuming $\mathcal{D}_c$, where:

$$\mathcal{S}_s(\tau) = \{S_k : \mathcal{D}_c \sqsubseteq S_k\}.$$

Situations describe sequences of events defined using $\mathcal{SDL}$, as described in Section 3. As soon as new sensory data are available (independently of the actual information source), the set $\mathcal{S}_s(\tau)$ contains all the patterns that have been detected to hold by the context assessment process. This information can be used to further driver the current system behaviour, for instance using a simple rule-based system or planning techniques [**9**].

5. An Example

5.1. Experimental Scenario. $\mathcal{SDL}$ has been used to model the behaviour of a team of service robots in a hospital environment, specifically taking into account issues related to *remote sensing* and *interaction with the environments*. The described approach has been experimented at the Polyclinic of Modena, Italy (see Figure 2). The robots are requested to transport – either according to a time schedule or on-demand – biologic waste and other material between areas located at different floors (see for instance areas in Figure 3) and – during the night – to perform patrolling activities. In order to reach storage sites, robots must navigate in complex and crowded buildings, where other mobile robots, hand-guided vehicles, stretchers and humans (both civilian and staff) can be often encountered. Each robot is provided with state-of-the-art techniques for self-localization [**10**], navigation and planning [**9**]. Furthermore, according to the Ubiquitous Robotics paradigm, a network of devices (i.e., sensors and actuators) has been installed. Nodes of the network are not limited to sense the environment: on the contrary, some of them are in charge of controlling automated appliances, such as intelligent elevators or material handling machines. In particular, the use of $\mathcal{SDL}$ to enforce context assessment capabilities is discussed with respect to a fundamental task, i.e., checking the feasibility of a mission to avoid the passage in crowded areas when the handled material is not safe.

5.2. Rationale. In order to complete a transportation mission, robots need to know both *in advance* and at run-time the *availability* of elevators (e.g., A and B in Figure 3), the *crowding* of areas to traverse (e.g., F, H and other areas $A_1,...,A_n$), the type of `material` being *handled* (in order to avoid contacts with food provisions and people) and the charge status of the `battery-pack`: in our set-up, this information is provided by proper information sources: an `elevator-system` agent running on the network device associated with elevators A and B continuously provides information about their *availability*; a `planning` agent [**9**] running on a networked workstation determines the sequence of areas $A_1,...,A_n$ to be visited in order to carry out the mission; proper algorithms (running on networked workstations) process camera images and passive infra red data (managed through sensor nodes in the network) in order to determine whether areas $A_1,...,A_n$ are *crowded* or not, whereas distributed sensors provide information to track the movements of other vehicles (through RFID tags) throughout the hospital; the integration with the hospital information system allows to access data about the *type* of `material` being handled and the privileged routes for other delivery activities; finally, the on-board agent `battery-manager` monitor batteries status, whereas other agents cooperate to perform obstacle avoidance, navigation and self-localization.

5.3. Situation Modelling. Assuming that F and H are – respectively – the initial and goal areas, and that $A_1,...,A_n$ are the areas to be visited by the robot in order to move from F to H, mission feasibility can be modelled as the *FeasibleMission Situation*, and in particular using the following collection of sentences (or "code") in $\mathcal{SDL}$:

(1) *PathExists* $\equiv$ *Connected(f, h)*;
(2) *UnsafeMaterial* $\equiv$ $\neg Safe(m)$;
(3) *FreePath* $\equiv$ $\neg Crowded(a_1, ..., a_n)$;
(4) *AvailableElevators* $\equiv$ *Available(a)* $\sqcup$ *Available(b)*;
(5) *ChargedBatteries* $\equiv$ $\delta^{f \rightarrow t}(Charged(battery\text{-}pack))$;

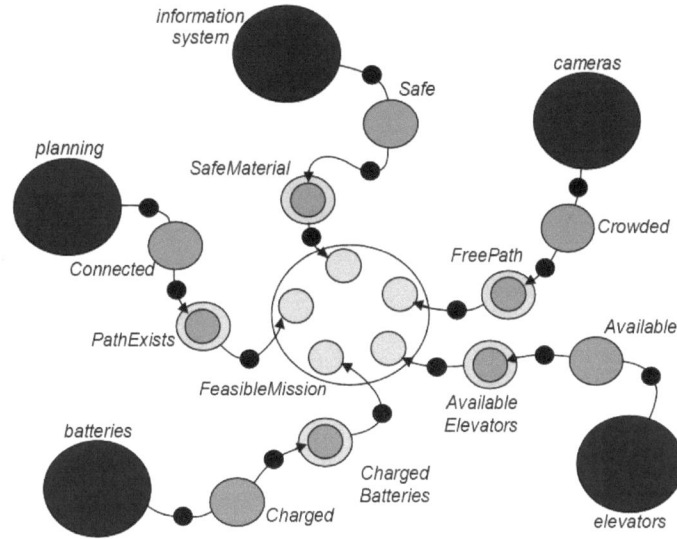

FIGURE 5. The *FeasibleMission* situation.

(6) $\sigma_1 \equiv$ *FeasibleMission* $\equiv$ *PathExists* $\sqcap$ *UnsafeMaterial* $\sqcap$ *FreePath* $\sqcap$ *AvailableElevators* $\sqcap$ *ChargedBatteries*;

Specifically, *FeasibleMission* is the actual *Situation* name, *PathExists*, *UnsafeMaterial*, *FreePath*, *AvailableElevators* and *ChargedBatteries* are *Contexts*, *Connected*, *Safe*, *Crowded*, *Available* and *Charged* are *Predicates*, whereas *f*, *h*, *m*, a_1, ..., a_n, *a*, *b* and *battery-pack* are variables which names correspond to environmental elements represented in Figure 3. Figure 5 shows a graphical representation of the symbolic structures involved in the definition of the *FeasibleMission Situation*.

5.4. Mapping FeasibleMission to the Ontology. Within the ontology, the $\mathcal{SDL}$ sentence σ_1 is mapped to a concept `FeasibleMission` subsumed by the general `Situation` concept, and defined as `PathExists` $\sqcap$ `UnsafeMaterial` $\sqcap$ `FreePath` $\sqcap$ `AvailableElevators` $\sqcap$ `ChargedBatteries`. Specifically, the concept `FeasibleMission` is made up of five `Context` concepts, which are connected through the *and* operator provided by the underlying ontology. Each `Context` is conventionally provided with a role `made-of` that is filled by a proper set of `Predicate` concepts. In this example, a one-to-one mapping exists between contexts and predicates: for instance, `ChargedBatteries` is associated with `Charged` using the description `ChargedBatteries` $\sqsubseteq$ `made-of.Charged`. `Predicate` concepts are related to their entities and grounded instances through their `arg` roles such that, for instance, `Charged` $\sqsubseteq$ `arg1.BatteryPack` $\sqcap$ `arg2.BatteryLevel`.

`Predicate` concepts have two child concepts that are used to describe their truth value. For instance, `Charged` subsumes `ChargedTrue` and `ChargedFalse`, which differs on the basis of the value of `BatteryLevel`. When this value is over a given threshold, the instance of `Charged` is subsumed by `ChargedTrue`, thereby contributing to `FeasibleMission` to hold. Finally, analogous mechanisms hold for the other `Context` instances as well.

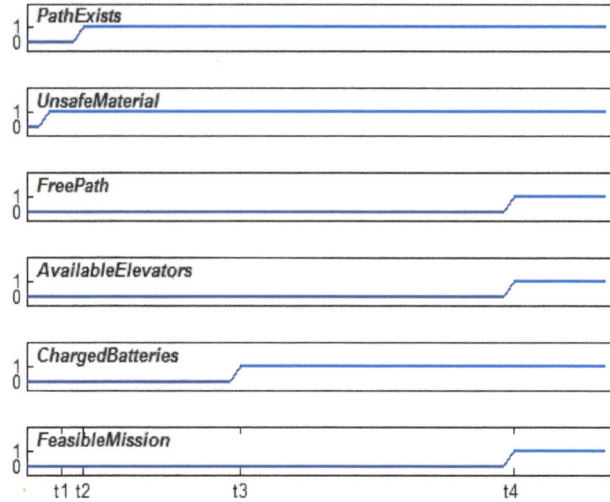

FIGURE 6. A variable assignment satisfying the *FeasibleMission* situation.

5.5. Situation Assessment. For a given interpretation $\mathcal{I}$ and time instant τ, if an α_τ is defined such that:

- *Connected*[f/F,g/G],
- ¬*Safe*[m/drug-1xZ],
- ¬*Crowded*[a_1/A$_1$,...,a_n/A$_n$],
- *Available*[a/A],
- ¬*Available*[b/B],
- *Charged*[*battery-pack*/battery-pack],

then the corresponding grounded sentence $\hat{\sigma}_1$ satisfies σ_1.

Figure 6 shows a real *truth value log* that satisfies this *Situation* (opportunely "stretched" to fit space limits): *FeasibleMission* holds at time instant t_4 when all the constituting instances of *Context* elements are satisfied. In particular, it can be noticed that, when the robot is requested to perform an actual mission, it first queries the hospital information system to retrieve the type of material to be delivered; then, an actual path to the goal area H is found. The mission can not start yet since batteries status is low. However, since the material drug-1xZ is considered unsafe, the mission is delayed until the areas to traverse A$_1$, ..., A$_n$ are free of people. In this case, *remote sensing* capabilities are exploited to mediate an implicit interaction (in this case, to avoid bare proximity) process between robots and humans.

6. Conclusions

In this Chapter, a formal language suitable to model context-aware behaviours for robots in intelligent environments has been formally presented and discussed. In order to ground actual descriptions, an example borrowed from the Ubiquitous Robotics paradigm has been deeply discussed, which is currently at the experimentation stage in a real-world scenario: specifically, remote sensing capabilities have been exploited to enforce implicit Human-Robot interaction processes. With respect to the three fundamental requirements that context models must adhere to, it is possible to conclude that:

- *effectiveness*: sensory data are mapped to symbolic representation that can be directly operated upon and used;
- *reusability*: since symbolic representations are maintained within an ontology in a hierarchical fashion, selected concepts and relationships can be differently composed to build different representations;
- *expressiveness*: since tractability of inference is a fundamental prerequisite, the system is limited to simple temporal relationships among events.

On the basis of current experimentation, practice suggests that using $\mathcal{SDL}$ it is possible to model a wide range of contexts and situations for a broader spectrum of artificial cognitive systems.

Bibliography

[1] Chen H., Finin T., Joshi A. An Ontology for Context Aware Pervasive Computing Environments. In *Proc. of the 18th Int.l Joint Conf. on Artificial Intelligence (IJCAI-03)*, Acapulco, Mexico, August 9-15, 2003.

[2] Cirillo M., Lazellotto F., Pecora F., Saffiotti A. Monitoring Domestic Activities with Temporal Constraints and Components. In *Proc. of the 5th Int.l Conf. on Intelligent Environments (IE'09)*, Barcelona, Spain, July 2009.

[3] Dey A.K. Understanding and Using Context. In *Personal and Ubiquitous Computing* **5**, 2001.

[4] Dourish P. What We Talk about when We Talk about Context. In *Personal and Ubiquitous Computing* **8**, 2004.

[5] Kim T.H., Choi S.H., Kim J.H. Incorporation of a Software Robot and a Mobile Robot Using a Middle Layer. In *IEEE Transactions on Systems, Man, and Cybernetics - Part C: Applications and Reviews* **37**(6), November 2007.

[6] Krummenacher R., Strang T. Ontology-Based Context-Modelling. In *Proc. of the 3rd Workshop on Context Awareness for Proactive Systems (CAPS'07)*, Guildford, United Kingdom, June 18–19, 2007.

[7] Loke S.W. Representing and Reasoning with Situations for Context-aware Pervasive Computing: a Logic Programming Perspective. In *Knowledge Engineering Review* **19**(3):213–233, 2005.

[8] Lonsdale P., Beale R. Towards a Dynamic Process Model of Context. In *Proc. of the 1st Workshop on Advanced Context Modelling, Reasoning and Management*, co-located with the *6th Int.l Conf. on Ubiquitous Computing (UbiComp'04)*, Nottingham, England, September 7–10, 2004.

[9] Mastrogiovanni F., Sgorbissa A., Zaccaria R. A System for Hierarchical Planning in Service Mobile Robotics. In *Proc. of the 8th Conf. on Intelligent Autonomous Systems (IAS-8)*, Amsterdam, The Netherlands, March 2004.

[10] Mastrogiovanni F., Sgorbissa A., Zaccaria R. The More the Better? A Discussion about Line Features for Self-Localization. In Proc. of the IEEE/RSJ 2007 Int.l Conf. on Intelligent Robots and Systems (IROS'07), San Diego, CA, November 2007.

[11] Mastrogiovanni F., Sgorbissa A., Zaccaria R. From Autonomous Robots to Artificial Ecosystems. In: Nakashima H., Aghajan H., and Augusto J.C. (Eds.) Handbook on Ambient Intelligence and Smart Environments. Springer-Verlag, Berlin Heidelberg, 2009.

[12] Mastrogiovanni F., Sgorbissa A., Zaccaria R. Context Assessment Strategies for Ubiquitous Robots. In *Proc. of the 2009 IEEE Int.l Conf. on Robotics and Automation (ICRA'09*, Kobe, Japan, May 2009.

[13] Sakamoto D., Hayashi K., Kanda T., Shiomi M., Koizumi S., Ishiguro H., Ogasawara T., Hagita N. Humanoid Robots as a Broadcasting Communication Medium on Open Public Spaces. In *International Journal of Social Robotics* **1**(2):157–169, April 2009.

[14] Shiomi M., Kanda T., Glas D., Satake S., Ishiguro H., Hagita N. Field Trial of Networked Social Robots in a Shopping Mall. In *Proc. of the 2009 IEEE/RSJ Int.l Conf. on Intelligent Robots and Systems (IROS'09)*, St. Louis, MO, USA, October 2009.

[15] Strang T., Linnhoff-Popien C., Frank K. CoOL: A Context Ontology Language to enable Contextual Interoperability. In *Proc. of 4th IFIP WG 6.1 Int.l Conf. on Distributed Applications and Interoperable Systems (DAIS2003)*, Paris, France, November 17–21, 2003.

[16] Strang T., Linnhoff-Popien C. A Context Modeling Survey. Proc. of the 6th Int.l Conf. on Ubiquitous Computing UbiComp2004. Nottingham, England, 2004.

[17] van Breemen A.J., Crucq K., Krose B.J.A., Nuttin M., Porta J.M., Deemester E. A User-Interface Robot for Ambient Intelligent Environments. Proc. of the First International Workshop on Advances in Service Robotics (ASER 2003), Bardolino, Italy, March 2003.

[18] Ye J., Coyle L., Dobson S., Nixon P. Ontology-based Models in Pervasive Computing Systems. In *Knowledge Engineering Review* **22**:315–347, 2007.

Send Orders of Reprints at reprints@benthamscience.net

Networking Humans, Robots and Environments, 2013, 59-75 59

CHAPTER 4

Self-Configurable Mobile Robot Swarms: Adaptive Triangular Mesh Generation

Geunho Lee and Nak Young Chong

School of Information Science

Japan Advanced Institute of Science and Technology

Ishikawa, Japan

ABSTRACT. We address the problem of dispersing a large number of autonomous mobile robots, for building wireless *ad hoc* sensor networks performing environmental monitoring and control. For this purpose, we propose the adaptive triangular mesh generation algorithm that enables robots to generate triangular meshes of various sizes, adapting to changing environmental conditions. A locally interacting, geometric technique allows each robot to generate a triangular mesh with its two neighbor robots. Specifically, we have assumed that robots are not allowed to have any identifiers, any pre-determined leaders or common coordinate systems, or any explicit communication. Under such minimal conditions, the positions of the robots were shown to converge to the desired distribution. This convergence was mathematically proven and also verified through extensive simulations. Our results indicate that the proposed algorithm can be applied to problems regarding the coverage of an area of interest by a swarm of mobile sensors.

1. Introduction

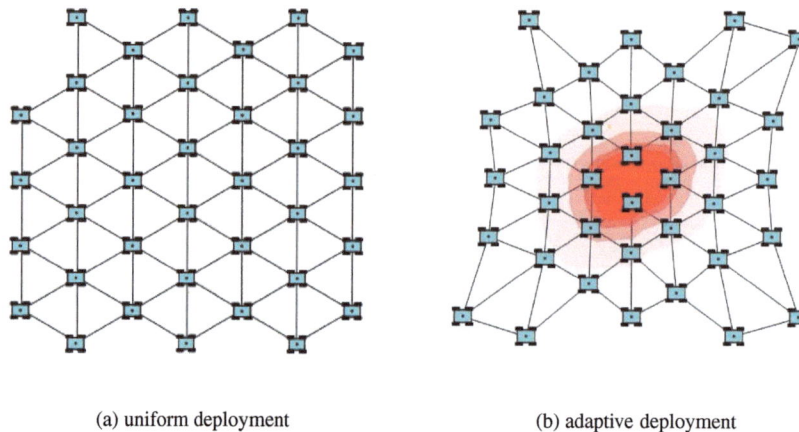

(a) uniform deployment   (b) adaptive deployment

FIGURE 1. Uniform vs. adaptive triangular meshes of mobile robot swarms

With the advance of wireless and mobile networking technologies, much attention has been paid to the use of large-scale swarms of simple, low-cost mobile robots for many applications [1] [2]. With the goal of deployment of robot swarms in real environments, researchers in swarm robotics have recently presented many fundamental coordination approaches such as self-configuration [9] [16], pattern formation [18] [19], flocking [20] [21], and consensus [22] [23]. Based on these coordination approaches, robot swarms are expected to perform a wide variety of real tasks, such as environmental or habitat monitoring [14] [17], exploration [24], search-and-rescue [25], odor localization [26]- [28], and so on.

In particular, for environmental or habitat monitoring, self-configuration of robot swarms requires a type of collective behavior that allows robots to disperse themselves in a certain area at a uniform spatial density. Thus, it is essential to properly coordinate the (relative) positions of robots, and this issue has been widely reported in literature [5]- [15]. Taking steps to further improve on those previous approaches, this work is aimed at presenting an algorithm to enable robot swarms to configure themselves adaptively in an area of interest with varying spatial densities. As illustrated in Fig. 1, robot swarms can explore an unknown area and detect and sense oil or chemical spills across the area. The contaminated area should be covered efficiently with mobile robots or sensors to investigate the degree and extent of contamination as quickly as possible and, if possible, prevent the possible expansion of the area. Therefore, in this paper, we address the problem of how to enable swarms of autonomous mobile robots to self-adjust their configuration or spatial density to fit local environmental conditions.

Based on our prior research on swarm configuration [16] [17], we propose an adaptive triangular mesh generation algorithm that enables robot swarms to explore an area, and to adjust the intervals between neighboring robots autonomously. The main objective is to provide robots with adaptive deployment capabilities to cover an area of interest more efficiently with variable triangular meshes according to sensed area conditions. This also can give us a more accurate picture of variations in the conditions of an area. In this chapter, the properties of the proposed algorithm are mathematically explained and the convergence is proven. We also demonstrated through extensive simulations that a large-scale swarm of robots can establish a triangular mesh network, adapting to varying degrees of connection. The results have been encouraging, and indicate that self-configurable robot swarms can be deployed for environmental or habitat monitoring.

The rest of this chapter is organized as follows. Section 2 gives a brief description on the state-of-the-art in the field of robot swarms. Section 3 presents the formal definitions of the adaptive triangular mesh generation problem. Section 4 describes our approach, its mathematical properties, and convergence at the equilibrium state. Section 5 summarizes the results of simulations. Section 6 explains our conclusions.

2. Background

Decentralized control for robot swarms can be broadly classified into global and local strategies, according to whether sensors have range limits. Global strategies [3] [4] [19] may provide fast, accurate, and efficient deployment, but are technically infeasible, and lack scalability as the number of robots increases. On the other hand, local strategies are mainly based on interactions between individual robots, inspired by nature. Local strategies can further be divided into biological emergence [5] [6], behavior-based [7], and virtual physics-based [8]- [15] approaches. Many of the behavior-based and virtual physics-based approaches use such physical phenomena as electric charges [8], gravitational forces [9], spring forces [10] [14] [15], potential fields [11], van der Waals forces [12], and other virtual models [13].

Robot swarm configurations achieved by the above-mentioned local interactions may result in lattice-type networks. These configurations offer high level coverage and multiple redundant connections, ensuring maximum reliability and flexibility from the standpoint of topology. Depending on whether there are interactions among all robots, the networks can be classified as fully or partially-connected topologies [30]. The fully-connected topologies have each robot interact with all other robots within a certain range simultaneously. Thus, those approaches might over-constrain individual robots, and frequently lead to deadlocks. However, using the partially-connected topology, robots interact selectively with other robots, but are connected to all robots. For example, robots may choose to exert force in a certain direction [14], where this selective interaction helps prevent them from being too tightly constrained. Therefore, robots may be able to achieve faster formation without deadlocks [15].

In our earlier work [**17**], we presented self-configuration of a robot swarm which enables a large number of robots to configure themselves into a 2-dimensional plane with geographic constraints. A locally-interacting geometric technique based on partially-connected topology provides a unique solution that allows robots to converge to uniform distribution by forming an equilateral triangle with two neighbors. By collecting this local behavior of each robot, a uniformly spaced swarm of robots was organized to cover an environment. Unlike previous works [**5**]- [**15**], our approach first was to construct uniformly spaced equilateral triangles conforming to the borders of an unknown area, when the robot sensors are subject to range and accuracy limitations. Second, an equilateral triangle lattice is built, with a partially-connected mesh topology. Among all the possible types of regular polygons, equilateral triangle lattices can reduce the computational burden, and are less influenced by other robots, due to the limited number of neighbors, and are highly scalable. The proposed local interaction is computationally efficient, since each robot utilizes only position information of two other robots. Our approach eliminates such major assumptions as robot identifiers, common coordinates, global orientation, and direct communication. More specifically, robots compute the target positions without requiring memories of past actions or states, helping to cope with transient errors.

3. Problem Statement

Definition and Notation. We consider a swarm of mobile robots, denoted as $r_1, \cdots, r_n$. It is assumed that all robots are within a swarm network configured by our previously proposed self-configuration method [**16**] [**17**]. Each robot autonomously moves on a 2-dimensional plane. They have no leader and no identifiers, do not share any common coordinate system, and do not retain any memory of past actions. Due to limited sensing range, they can detect the position of other robots only within a certain distance. In addition, robots do not communicate explicitly with other robots.

Let us consider a *robot* r_i with local coordinates $\vec{r}_{x,i}$ and $\vec{r}_{y,i}$, as illustrated in Fig. 2-(a). Here, $\vec{r}_{y,i}$ defines the vertical axis of r_i's coordinate system as its heading direction. It is straightforward to determine the horizontal axis $\vec{r}_{x,i}$ by rotating the vertical axis 90 degrees counterclockwise. The *position* of r_i is denoted as p_i. Note that p_i is $(0,0)$ with respect to r_i's local coordinates. The *distance* between p_i of r_i and p_j of another robot r_j is denoted as $dist(p_i, p_j)$. We define a *uniform distance* d_u, the predefined desired interval between r_i and r_j. As shown in Fig. 2-(b), r_i detects the positions p_j, p_k, and p_l of other robots located within its sensing boundary *SB*, yielding an *observation set* of the positions O_i (=$\{p_j, p_k, p_l\}$) with respect to its local coordinates. Next, r_i can select two robots r_{s1} and r_{s2} within r_i's *SB*, which we call the *neighbors* of r_i, and denote their positions, $\{p_{s1}, p_{s2}\}$, as N_i. Given p_i and N_i in Fig. 2-(c), the *triangular configuration*, denoted by $\mathbb{T}_i$, is defined as a set of three distinct positions $\{p_i, p_{s1}, p_{s2}\}$.

As mentioned above, if robots detect an event such as an oil or chemical spill within the self-configured network, they attempt to cooperate with each other to cover the area as efficiently as possible. The gradient in contamination density across the area forces robots to adjust the intervals between neighboring robots. For a certain point p_i occupied by r_i as presented in Fig. 2-(d), the densities are expressed by k_i ranging between $0 < k_i \leq 1$, where $k_i = 0$ represents the maximum density and $k_i = 1$ corresponds to zero density. Moreover, it is assumed that each robot can detect and measure densities for positions occupied by other robots within *SB*, yielding a set of the densities K_i corresponding to individual positions in O_i.

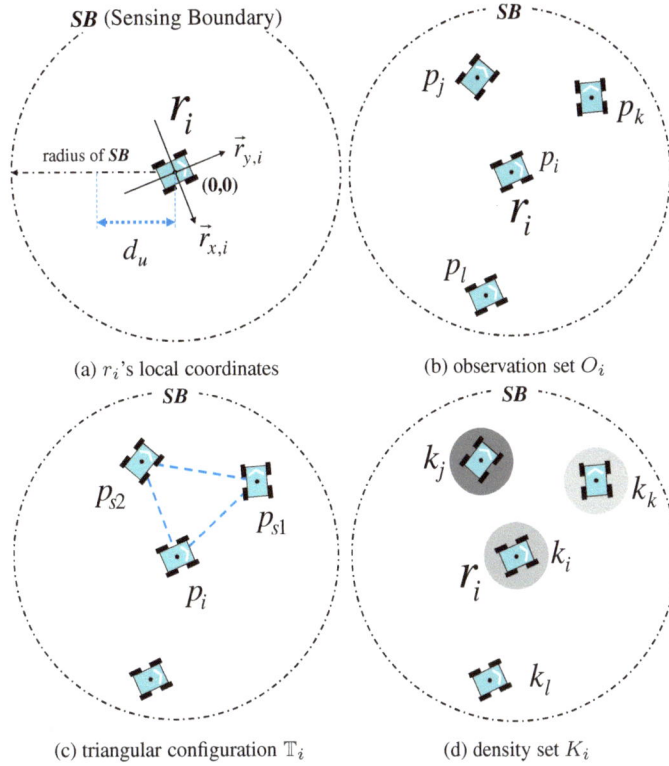

(a) r_i's local coordinates

(b) observation set O_i

(c) triangular configuration $\mathbb{T}_i$

(d) density set K_i

FIGURE 2. Illustration of definitions and notations frequently used in this chapter

Problem Definition. Now, we formally address the ADAPTIVE TRIANGULAR MESH GENERATION problem as follows.

Given a swarm of mobile robots self-configured in a 2-dimensional plane, how can the robots form triangular mesh patterns of various sizes adapting to varying environmental conditions?

Our approach to the above problem enables robots to disperse themselves into equilateral triangular patterns of various sizes, according to changes in environmental conditions within an area of interest. The basic concept behind this approach is to enable robots to form triangular configurations while changing their suitable distances depending on locally-sensed information. In other words, given a uniform distance d_u, three neighboring robots configure an equilateral triangle with a side length proportional to the local measurement data density.

4. Solution Approach

Regarding our solution approach, this section explains the generation of triangular meshes adapting to the measurement data densities in the areas occupied by three neighboring robots, provides two important properties of the approach, and shows the convergence into a desired stable configuration using Lyapunov's theory.

ALGORITHM-1   ADAPTIVE TRIANGULAR MESH GENERATION

Function $\varphi_{triangular}(O_i, K_i)$

1   $p_{s1} := \min\limits_{p \in O_i - \{p_i\}} [dist(p_i, p)]$

2   $p_{s2} := \min\limits_{p \in O_i - \{p_i, p_{s1}\}} [dist(p_{s1}, p) + dist(p, p_i)]$

3   $\phi :=$ angle between $\overline{p_{s1}p_{s2}}$ and r_i's local horizontal axis

4   $p_{ct} := (p_{ct,x}, p_{ct,y})$ // *centroid of* $\mathbb{T}_i (= \{p_{s1}, p_{s2}, p_i\})$

5   $k_a := \frac{k_i + k_{s1} + k_{s2}}{3}$ // average contamination density

6   $d_a := k_a \times \frac{d_u}{\sqrt{3}}$ // desired interval from p_{ct}

7   $p_{ti,x} := p_{ct,x} + d_a \cos(\phi + \frac{\pi}{2})$

8   $p_{ti,y} := p_{ct,y} + d_a \sin(\phi + \frac{\pi}{2})$

9   $p_{ti} := (p_{ti,x}, p_{ti,y})$ // next target point

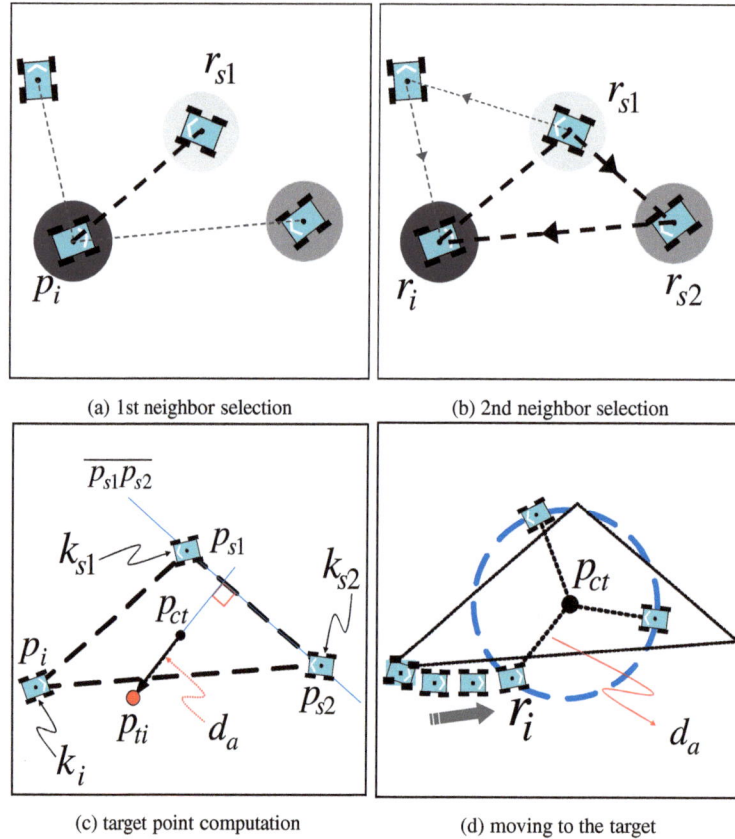

(a) 1st neighbor selection

(b) 2nd neighbor selection

(c) target point computation

(d) moving to the target

FIGURE 3. Illustration of ALGORITHM-1

Algorithm Description. Here, we describe the adaptive triangular mesh generation algorithm. As presented in ALGORITHM-1, the algorithm consists of a function $\varphi_{triangular}$ whose arguments are O_i and K_i at each activation. Each time, r_i first observes other robots within O_i to select the closest neighbor r_{s1} as illustrated in Fig. 3-(a). If there exist two or more candidates for r_{s1}, r_i determines r_{s1} according to the high contamination density. Secondly, as illustrated in Fig. 3-(b), the second neighbor r_{s2} within O_i is selected, such that the total distance from the position p_{s1} of r_{s1} to p_i passing through p_{s2} is minimized. Similarly, if there are two or more candidates for r_{s2}, r_i selects r_{s2} occupied in the area with higher density of contamination. Then, as illustrated in Fig. 3-(c), r_i measures the angle ϕ between the line $\overline{p_{s1}p_{s2}}$ connecting two neighbors and the horizontal axis of the observing r_i's coordinate system. Thirdly, the centroid p_{ct} in $\mathbb{T}_i$ ($\triangle p_i p_{s1} p_{s2}$) is computed. Moreover, based on k_i, k_{s1}, and k_{s2} at each position occupied by r_i, r_{s1}, and r_{s2}, r_i finds the local average density k_a in $\mathbb{T}_i$ through the computation of $(k_i + k_{s1} + k_{s2})/3$. Then, from p_{ct}, r_i calculates an appropriate interval as follows: $d_a = k_a \times d_u/\sqrt{3}$. Utilizing d_a and ϕ, r_i calculates its target point $p_{ti} = (p_{ti,x}, p_{ti,y})$ located on the same line as the previously calculated interval from p_{ct} and perpendicular to $\overline{p_{s1}p_{s2}}$. Finally, r_i moves toward p_{ti} as illustrated in Fig. 3-(d). By repeating this process, r_i can form a triangular mesh depending on the contamination densities.

Mathematical Properties. Let's consider a triangle (whose centroid is p_{ct}) $\triangle p_i p_{s1} p_{s2}$ (or $\mathbb{T}_i$) configured from the three positions occupied by r_i, r_{s1}, and r_{s2}. By ALGORITHM-1 above, at time t, r_i in $\mathbb{T}_i(t)$ finds the next target point p_{ti} of which the line segment $\overline{p_{ct}p_{ti}}$ is $k_a d_u/\sqrt{3}$ in length and is perpendicular to $\overline{p_{s1}p_{s2}}$ in Fig. 3-(c). In other words, at $t+1$, the height of $\triangle p_{ti} p_{s1} p_{s2}$ is the straight line through p_{ti} and perpendicular to $\overline{p_{s1}p_{s2}}$. Similarly, since r_{s1} and r_{s2} also execute the same algorithm, it is easily seen that p_{ct} at t is the orthocenter H at $t+1$.

In Fig. 4, we denote p_i, p_{s1}, and p_{s2} for simplicity as A, B, and C, respectively. The lengths of lines $\overline{AB}$, $\overline{AC}$, and $\overline{BC}$ are denoted as c, b, and a, respectively. The points P, Q, and R are the foot of the perpendiculars from the vertices C, B, and A to the vectors $\overrightarrow{AB}$, $\overrightarrow{AC}$, and $\overrightarrow{BC}$, respectively. The angles $\angle CAB$, $\angle ABC$, and $\angle BCA$ are denoted as α, β, and γ, respectively. Moreover, H is the orthocenter of $\triangle ABC$. Moreover, H is the orthocenter of $\triangle ABC$ (Since p_{ct} and H exist in the same location under ALGORITHM-1, we also use only H instead of p_{ct}).

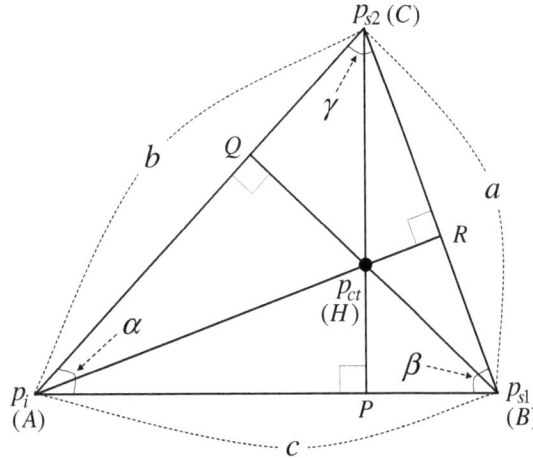

FIGURE 4. Illustrating the derivation of r_i's position vector toward H

Since $\overrightarrow{AB}$ and $\overrightarrow{AC}$ are linearly independent, $\overrightarrow{AH}$ can be defined as

(1)
$$\overrightarrow{AH} = x\overrightarrow{AB} + y\overrightarrow{AC},$$

where x and y are scaling coefficients. Since we can easily see that $\overrightarrow{AP} = \frac{b\cos\alpha}{c}\overrightarrow{AB}$, the following relation holds:

$$\overrightarrow{PH} = \overrightarrow{AH} - \overrightarrow{AP} = (x - \frac{b\cos\alpha}{c})\overrightarrow{AB} + y\overrightarrow{AC}.$$

Thus, the inner product between $\overrightarrow{PH}$ and $\overrightarrow{AB}$ can be expressed as follows:

(2)
$$\overrightarrow{PH} \cdot \overrightarrow{AB} = (x - \frac{b\cos\alpha}{c})c^2 + (bc\cos\alpha)y = 0.$$

Similarly, the following equation holds:

(3)
$$\overrightarrow{QH} \cdot \overrightarrow{AC} = (bc\cos\alpha)x + (y - \frac{c\cos\alpha}{b})b^2 = 0.$$

Now, using (2) and (3), the following simultaneous equations can be obtained:

(4)
$$cx - (b\cos\alpha)y = b\cos\alpha$$
$$(c\cos\alpha)x + by = c\cos\alpha.$$

By solving (4), we can obtain the coefficient x as follows:

$$x = \frac{b\cos\alpha(b - c\cos\alpha)}{bc\sin^2\alpha}.$$

Using the cosine formula ($b = c\cos\alpha + a\cos\gamma$), x is expressed as follows:

$$x = \frac{a\cos\alpha\cos\gamma}{c\sin^2\alpha}.$$

In addition, by utilizing the sine formula ($\frac{a}{\sin\alpha} = \frac{c}{\sin\gamma}$), x is rewritten as the following equation:

$$x = \frac{\cos\alpha\cos\gamma}{\sin\alpha\sin\gamma}.$$

Thus, if we do not consider the case of a right triangle, it is straightforward to rewrite x as the following equation:

(5)
$$x = \frac{1}{\tan\alpha\tan\gamma} = \frac{\tan\beta}{\tan\alpha\tan\beta\tan\gamma}.$$

Similarly, the coefficient y can be represented as follows:

(6)
$$y = \frac{\tan\gamma}{\tan\alpha\tan\beta\tan\gamma}.$$

Using the addition theorems of the trigonometric function, we can obtain the following result:

$$\tan\alpha\tan\beta\tan\gamma = \tan\alpha + \tan\beta + \tan\gamma.$$

Thus, (5) and (6) are rewritten as follows:

(7)
$$x = \frac{\tan\beta}{\tan\alpha + \tan\beta + \tan\gamma}$$
$$y = \frac{\tan\gamma}{\tan\alpha + \tan\beta + \tan\gamma}.$$

With respect to a reference point O, (1) can be rewritten as follows:
$$\overrightarrow{AH} = \overrightarrow{OH} - \overrightarrow{OA} = x(\overrightarrow{OB} - \overrightarrow{OA}) + y(\overrightarrow{OC} - \overrightarrow{OA}).$$

Now $\overrightarrow{OH}$ can be represented in the following form:

(8) $$\overrightarrow{OH} = (1 - x - y)\overrightarrow{OA} + x\overrightarrow{OB} + y\overrightarrow{OC}.$$

Substituting (7) into (8), finally, $\overrightarrow{OH}$ is given by

(9) $$\overrightarrow{OH} = \frac{\tan\alpha\,\overrightarrow{OA} + \tan\beta\,\overrightarrow{OB} + \tan\gamma\,\overrightarrow{OC}}{\tan\alpha + \tan\beta + \tan\gamma}.$$

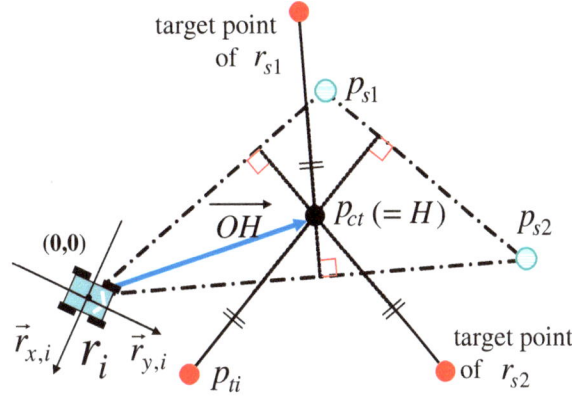

FIGURE 5. Illustrating two important properties: p_{ct} and position vector

Under the adaptive triangular mesh generation algorithm, there are two mathematical properties. First, p_{ti} is determined with p_{ct} in $\mathbb{T}_i(t)$ at t used as H at $t + 1$. Moreover, r_i uses p_{ct} at t as a basis to generate $\mathbb{T}_i(t + 1)$ with d_r from p_{ct} to p_{ti} at $t + 1$ (see Fig. 3-(c)). Secondly, since p_{ct} and H between t and $t + 1$ remain unchanged, $\overrightarrow{OH}$ in (9) is the position vector toward p_{ct} with respect to the origin of r_i's local coordinates as illustrated in Fig. 5.

Motion Control. Under the adaptive triangular mesh generation algorithm, three neighboring robots attempt to cooperatively configure themselves into an equilateral triangle, adapting to the density of contamination. As presented in Fig. 6, let's consider the circumscribed an equilateral triangle $\triangle p_{ti}p_{s1}p_{s2}$ configured from three positions occupied by r_i, r_{s1}, and r_{s2} where the center is p_{ct} and the radius is d_a $(= k_a \times \frac{d_u}{\sqrt{3}})$ in length. From the triangular configuration and the two mathematical properties above, we design the motion of r_i by controlling the distance d_i from p_{ct} and the internal angle θ_i between $\overline{p_{ct}p_{ti}}$ and $\overline{p_{ct}p_{s2}}$.

First, d_i in Fig. 6 is controlled by the following equation:

(10) $$\dot{d}_i(t) = -a(d_i(t) - d_a),$$

where a is a positive constant. Indeed, the solution of (10) is obtained as follows:

$$d_i(t) = |d_i(0)|e^{-at} + d_a.$$

The solution converges exponentially to d_a as t approaches infinity. Secondly, θ_i is controlled by the following equation:

(11) $$\dot{\theta}_i(t) = k(\theta_{s1}(t) + \theta_{s2}(t) - 2\theta_i(t)),$$

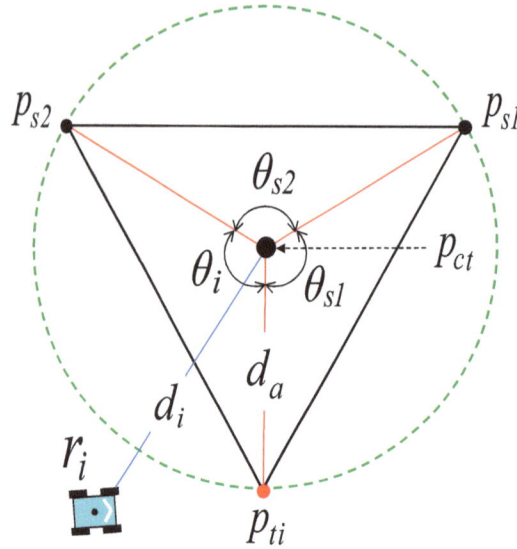

FIGURE 6. Motion control of a robot by the algorithm

where k is a positive number. Using the features of a triangle whose total external angles is 2π, (11) can be rewritten as

$$(12) \qquad \dot{\theta}_i(t) = k'(\frac{2}{3}\pi - \theta_i(t)),$$

where k' is $3k$. Similarly, the solution of (12) is obtained as follows:

$$\theta_i(t) = |\theta_i(0)|e^{-k't} + \frac{2}{3}\pi.$$

The soultion converges exponentially to $\frac{2}{3}\pi$ as t approaches infinity. Note that (10) and (12) imply that the trajectory of r_i converges to an equilibrium state $[d_a \ \frac{2}{3}\pi]^T$. From Fig. 6, this also implies that the equilibrium for θ_i is defined as $\theta_i = \theta_{s1}$ since $\triangle p_{ti}p_{ct}p_{s1}$ and $\triangle p_{ti}p_{ct}p_{s2}$ are eventually congruent [**16**]. In order to show the convergence into the state $[d_i(t) \ \theta_i(t)]^T$, we will take advantage of stability based on Lyapunov's theory [**31**]. The convergence into the desired stable configuration is one that minimizes the energy level of the scalar function. Consider the following scalar function:

$$(13) \qquad f_i(d_i, \theta_i, \theta_{s1}) = \frac{1}{2}(d_i - d_a)^2 + \frac{1}{2}(\theta_{s1} - \theta_i)^2.$$

This scalar function is always positive definite except when $d_i \neq d_a$ and $\theta_i \neq \theta_{s1}$. The derivative of the scalar function is given as follows:

$$\dot{f}_i = -(d_i - d_a)^2 - (\theta_{s1} - \theta_i)^2,$$

which is negative definite. The scalar function is radially unbounded, since it tends to infinity as $||[d_i(t) \ \theta_i(t)]^T|| \to \infty$. Therefore, the equilibrium state is asymptotically stable, implying that r_i reaches a vertex of the stable triangle.

Now we show the convergence of the algorithm for n robots. The n-order scalar function $\mathbf{F}$ is defined as

$$(14) \qquad \mathbf{F} = \sum_{i=1}^{n} f_i(d_i, \theta_i, \theta_{s1}).$$

It is straightforward to verify that $\mathbf{F}$ is positive definite and $\dot{\mathbf{F}}$ is negative definite. $\mathbf{F}$ is radially unbounded, since it tends to infinity as t approaches infinity. Consequently, n robots move toward the equilibrium state.

5. Simulation Results and Discussion

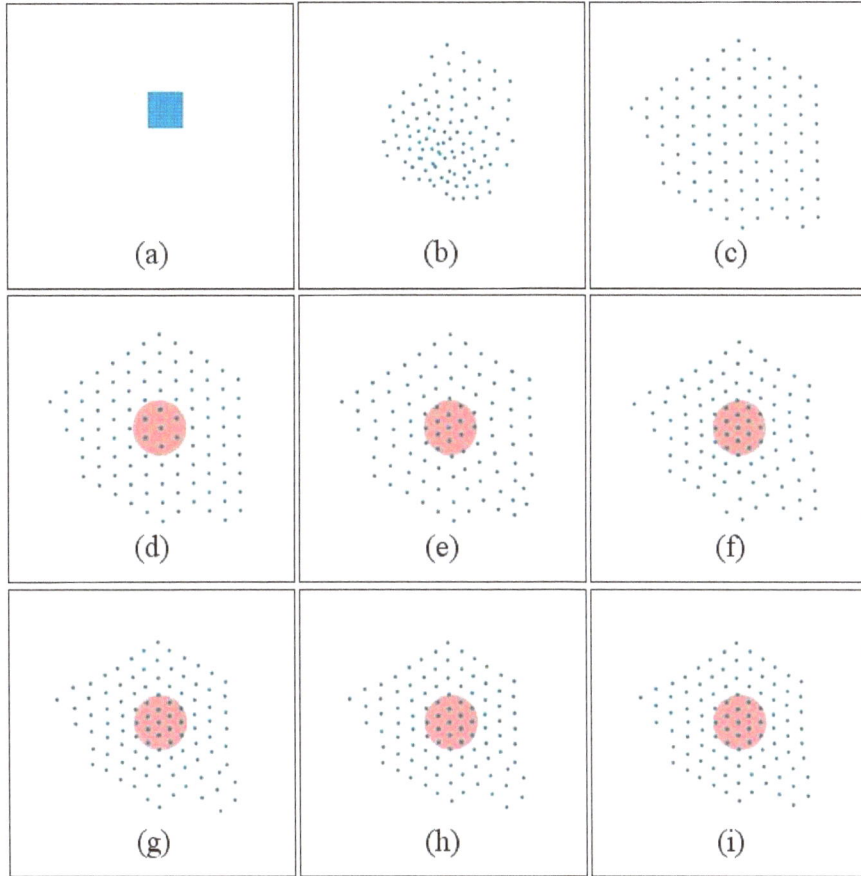

FIGURE 7. Simulation result for the whole deployment of 100 robots ((a)~(c): uniform configuration [**16**], (d)~(i): adaptive configuration with increasing center density of 0.7)

In this section, we describe simulations of filling an area of interest with a swarm of robots in order to show the validity of our proposed algorithm. In these simulations, we represent the area of varying contamination density as a colored circle that will be sparsely or densely populated with robots.

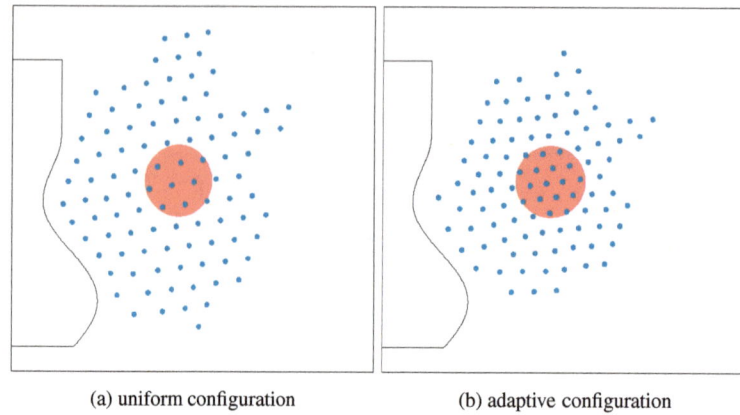

(a) uniform configuration   (b) adaptive configuration

FIGURE 8. Adaptive triangular mesh generation with geographic borders

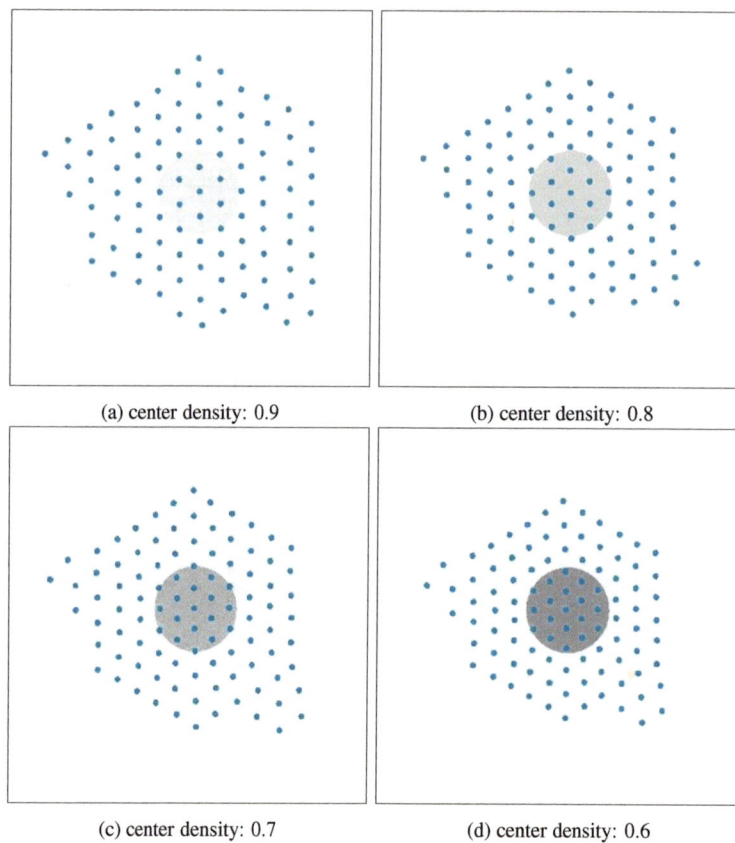

(a) center density: 0.9   (b) center density: 0.8

(c) center density: 0.7   (d) center density: 0.6

FIGURE 9. Simulation results according to varying contamination densities

First, in order to assist in understanding the self-configuration of robot swarms from an initial distribution, Fig. 7 presents snapshots for the simulation result by 100 robots. From the initial distribution in Fig. 7-(a), the robots configured themselves in the 2-dimensional plane (see Fig. 7-(c)). After constructing an equilateral triangle network, we investigated how they adapt their triangular mesh network according to the assigned contamination densities. The snapshots from Fig. 7-(d) to (i) show that robots could generate adaptive triangular meshes based on k_a and d_a. If the series of snapshots are carefully observed, the number of robots within the area of interest, represented by a red circle, increases. In detail, it is initially observed in Fig. 7-(d) that there are 7 robots within the area. By executing the adaptive triangular mesh generation algorithm repeatedly, the number of robots located in that area rises to 17 robots in the final distribution (see Fig. 7-(i)). Compared with Fig. 7-(c), the overall size of the final distribution was reduced. Similarly, Fig. 8 shows the result from another simulation with geographical borders [17].

Secondly, Fig. 9 shows the results from simulations with four different degrees of center density. In the figure, the number of robots within the circle increases in proportion to the degree of density. Similarly, the size of the swarm in its final converged shape varies according to the center density. Higher densities forced the robots to decrease the intervals between neighboring robots.

Thirdly, we performed simulations for multiple varying densities in a single swarm. Figs. 10-(a) and (b) show the results for two identical center densities. Although the swarm was spilt into two smaller groups, robots could adaptively configure themselves near the area. Figs. 10-(c) and (d) present the results when the center contamination densities vary: 0.7 on the left hand side and 0.9 on the right hand side. The higher center contamination density area is more densely populated. In Figs. 10-(e) and (f), the simulation was performed for three different center contamination densities in a swarm: 0.9, 0.7, and 0.8 from left to right. Robots could adapt the interval between each other according to the varying contamination densities, which is similar to previous simulations. Thus the overall size of the swarm also varies accordingly.

Three main features highlight our adaptive triangular mesh generation as follows. First, the approach enables the swarm of robots to deploy themselves while adapting to contamination densities. More specially, we proposed motion control to form equilateral triangles with a appropriate intervals depending on the density. Secondly, the triangle lattice is built based on a partially-connected mesh topology since each robot locally interacts with its selected neighbors. Among all the possible types of regular polygons, the triangle element is easy to construct and highly scalable as the number of robots increases. Thirdly, our approach eliminates major assumptions such as robot identifiers, common coordinates, global orientation, specific leaders, and direct communication. Robots calculate their target position without having to remember past actions or states, which makes it easier to cope with transient error.

We believe that our algorithm works well under real world conditions, but several issues remain to be addressed. Our approach relies on the assumption that robots can sense the positions of neighboring robots and contamination densities within *SB*. Practically speaking, it is difficult to precisely measure the positions of other robots using infrared [9] [29] or sonar sensors [19], and to detect the contamination densities. When direct communications are employed, it is possible to exchange information about densities. So, as our future work, (to further facilitate implementation of the proposed method in a real environment,) robot swarms exchanging information are expected to be effectively applied to such deployment. Robots, however, still require *a priori* knowledge, such as individual identifiers or global coordinates. Direct communication may also involve difficulties such as limited bandwidth, range, and interference. These important engineering issues are left for future work.

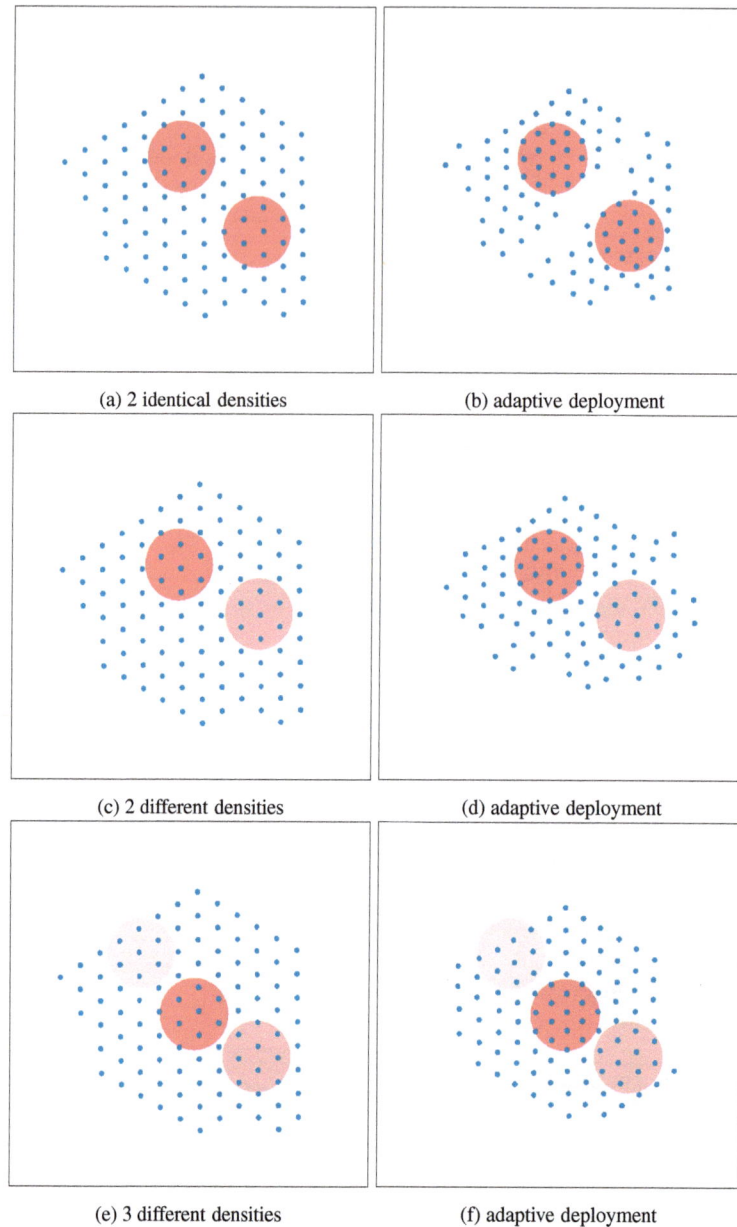

(a) 2 identical densities

(b) adaptive deployment

(c) 2 different densities

(d) adaptive deployment

(e) 3 different densities

(f) adaptive deployment

FIGURE 10. Simulation results for varying local contamination densities

6. Conclusions

The adaptive triangular mesh generation problem was addressed to disperse a swarm of mobile robots which can adapt to the degree of contamination in a target area. There were several major assumptions underlying our proposed approach to this problem: no robot identifiers, no common coordinates or global orientation, and no direct communication. Robots computed their target positions without requiring memories of past actions or states. Under such conditions, the proposed adaptive mesh generation algorithm enables a large-scale swarm of robots to configure themselves into triangular patterns, while changing the uniform interval according to the contamination densities that can be detected by sensors. We took advantage of the fact that, among all the possible types of n-polygons, the triangle is highly scalable, and less influenced by the number of neighboring robots. To form the triangular pattern, robots were allowed to interact with only two selected neighbors at each time. By collecting such a local behavior of each robot, a swarm of robots arranged in triangular meshes was self-configured into the area of varying degrees of the local density. The properties of the algorithm were shown mathematically, and also verified through extensive simulations. Finally, we expect that the proposed approach can be used as a simple and effective way to deploy mobile sensor networks for coverage of unknown areas of interest.

Bibliography

[1] Choset, H., "Coverage for robotics-a survey of recent results". Annals of Mathematics and Artificial Intelligence **31**(1-4):113-126 (2001).

[2] Cortes, J., Martinez, S., Karatas, T., and Bullo, F., "Coverage control for mobile sensing networks". IEEE Transactions on Robotics and Automation **20**(2):243-255 (2004).

[3] Suzuki, I. and Yamashita, M., "Distributed anonymous mobile robots: formation of geometric patterns". SIAM Journal of Computing, **28**(4):1347-1363 (1999).

[4] Cao, Z., Tan, M., Wang, S., Fan, Y., and Zhang, B., "The optimization research of formation control for multiple mobile robots". Proc. 4th World Cong. Intelligent Control and Automation, pp.1270-1274 (2002).

[5] Ikemoto, Y., Hasegawa, Y., Fukuda, T., and Matsuda, K., "Graduated spatial pattern formation of robot group". Information Sciences **171**(4):431-445 (2005).

[6] Shimizu, M., Mori, T., and Ishiguro, A., "A development of a modular robot that enables adaptive reconfiguration". Proc. IEEE/RSJ Int. Conf. Intelligent Robots and Systems, pp.174-179 (2006).

[7] Balch, T. and Hybinette, M., "Social potentials for scalable multi-robot formations". Proc. IEEE Int. Conf. Robotics and Automation, pp.73-80 (2000).

[8] Howard, A., Mataric, M.J., and Sukhatme, G.S., "Mobile sensor network deployment using potential fields: a distributed, scalable solution to the area coverage problem". Proc. 6th Int. Symp. Distributed Autonomous Robotic Systems, pp.299-308 (2002).

[9] Spears, W., Spears, D., Hamann, J., and Heil, R., "Distributed, physics-based control of swarms of vehicles". Autonomous Robots **17**(2-3):137-162 (2004).

[10] Fujibayashi, K., Murata, S., Sugawara, K., and Yamamura, M., "Self-organizing formation algorithm for active elements". Proc. 21st IEEE Symp. Reliable Distributed Systems, pp.416-421 (2002).

[11] Reif, J. and Wang, H., "Social potential fields: a distributed behavioral control for autonomous robots". Robotics and Autonomous Systems **27**(3):171-194 (1999).

[12] Zheng, Y.F. and Chen, W., "Mobile robot team forming for crystallization of protein". Autononomous Robots **23**(1):69-78 (2007).

[13] Martison, E. and Payton, D. "Lattice formation in mobile autonomous sensor arrays" in *Swarm Robotics* (*LNCS*). Sahin, E. and Spears, W.M. (eds.), **3342**:98-111, Springer, (2005).

[14] McLurkin, J. and Smith, J., "Distributed algorithms for dispersion in indoor environments using a swarm of autonomous mobile robots". Proc. 7th Int. Symp. Distributed Autonomous Robotic Systems, pp.831-840 (2004).

[15] Shucker, B., Murphey, T.D., and Bennett, J.K., "Convergence-preserving switching for topology-dependent decentralized systems". IEEE Transactions on Robotics **24**(6):1405-1415 (2008).

[16] Lee, G. and Chong, N.Y., "A geometric approach to deploying robot swarms". Annals of Mathematics and Artificial Intelligence **52**(2-4):257-280 (2008).

[17] Lee, G. and Chong, N.Y., "Self-configurable mobile robot swarms with hole repair capability". Proc. IEEE/RSJ Int. Conf. Intelligent Robots and Systems, pp.1403-1408 (2008).

[18] Fredslund, J. and Mataric, M.J., "A general algorithm for robot formations using local sensing and minimal communication". IEEE Transactions on Robotics and Automation **18**(5):837-846 (2002).

[19] Lee, G. and Chong, N.Y., "Decentralized formation control for small-scale robot teams with anonymity". Mechatronics **19**(1):85-105 (2009).

[20] Folino, G. and Spezzano, G., "An adaptive flocking algorithm for spatial clustering" in *Parallel Problem Solving from Nature - PPSN VII* (*LNCS*). Goos, G., Hartmanis, J., and Leeuwen, J.V. (eds.), **2439**:924-933, Springer,(2002).

[21] Lee, G. and Chong, N.Y., "Adaptive flocking of robot swarms: algorithms and properties". IEICE Transactions on Communications **E91-B**(9):2848-2855 (2008).

[22] Fax, J.A. and Murray, R.M., "Information flow and cooperative control of vehicle formations". IEEE Transactions on Automatic Control **49**(9):1465-1476 (2004).

[23] Olfati-Saber, R. and Murray, R.M., "Consensus problems in networks of agents with switching topology and time-delays". IEEE Transactions on Automatic Control **49**(9):1520-1533 (2004).

[24] Burgard, W., Moors, M., Stachniss C., and Schneider, F., "Coordinated multi-robot exploration". IEEE Transactions on Robotics and Automation **20**(3):120-145 (2005).

[25] Mondada, F., Pettinaro, G.C., Guignard, A., Kwee, I., Floreano, D., Deneubourg, J.-L., Nolfi, S., Gambardella, L.M., and Dorigo, M., "Swarm-bot: a new distributed robotic concept". Autonomous Robots **17**(2-3):193-221 (2004).

[26] Jatmiko, W., Sekiyama, K., and Fukuda, T., "A particle swarm-based mobile sensor network for odor source localization in a dynamic environment" in *Distributed Autonomous Robotic Systems 7*. Gini, M. and Voyles, R. (eds.), Springer Japan, pp.71-80 (2007).

[27] Jung, B. and Sukhatme, G.S., "Tracking targets using multiple robots: the effect of environment occlusion". Autonomous Robots **13**(3):191-205 (2002).

[28] Krishnanand, K.N., Amruth, P., and Guruprasad, M.H., "Glowworm-inspired robot swarm for simultaneous taxis towards multiple radiation sources". Proc. IEEE Int. Conf. Robotics and Automation, pp.958- 963 (2006).

[29] Lee, G., Yoon, S., Chong, N.Y., and Christensen, H., "Self-configuring robot swarms with dual rotating infrared sensors". Proc. IEEE/RSJ Int. Conf. Intelligent Robots and Systems, pp.4357-4362 (2009).

[30] Ghosh, S., Basu, K., and Das, S.K., "An architecture for next-generation radio access networks". IEEE Network **19**(5):35-42 (2005).

[31] Slotine, J.E. and Li, W., *Applied nonlinear control*. Prentice-Hall (1991).

Send Orders of Reprints at reprints@benthamscience.net

CHAPTER 5

Experimental Validation of Multi-Agent Coordination by Decentralized Estimation and Control

Michael Hwang, Matthew L. Elwin, Peng Yang, Randy A. Freeman, and Kevin M. Lynch

Northwestern Institute on Complex Systems (NICO) and Departments of Mechanical Engineering and Electrical Engineering and Computer Science Northwestern University Evanston, IL 60208, USA

ABSTRACT. In previous work, we developed a decentralized framework for formation control by mobile robots. Each robot simultaneously estimates the current swarm shape, by local information diffusion in a communication network, while controlling its own motion to drive the swarm to the desired shape. These continuous-time estimation and control laws result in provably correct behavior, but they make unrealistic assumptions for implementation on actual robots.

This paper describes the first experimental implementation of decentralized estimation and control based on information diffusion. We develop the hardware, software, and communication protocols that adapt the theoretical approach to an actual implementation. Experiments demonstrate that the robots successfully estimate the first and second inertial moments of the swarm, using scalable local communication, while simultaneously moving to control these moments.

1. Introduction

We are pursuing a framework for self-organizing robot systems based on decentralized estimation and control. The goal is to "compile" desired group behaviors into local communication, estimation, and control laws running on individual robots. For properly designed control laws, the interactions of individual robots result in the desired group behavior, without any centralized control.

A key to this centralized-to-decentralized compilation process is the ability of each robot to estimate the global performance of the group based on local sensing and local communication in a time-varying communication network. For scalability, we require that the communication, computation, and storage requirements for each robot be independent of the total number of robots in the swarm. For robustness, we require that no robot be indispensable. In fact, we typically assume that each robot is identical.

To achieve robust and scalable global performance estimation, we and others have developed *dynamic average consensus estimators* [2, 5, 7, 11, 12, 17]. These estimators allow each robot to estimate the average of time-varying inputs from across the whole swarm. Estimates spread through the network in a process similar to heat diffusion, and we sometimes refer to the process as *information diffusion*.

Many functions of sensor inputs can be estimated using average consensus estimators, as we can first apply a transformation to the sensor inputs, pass them through the averaging process, and then apply a transformation to the output to obtain the desired result. We have shown that average consensus estimators allow the compilation of global objectives into local controllers for tasks such as formation control [6,16], cooperative target tracking [15], estimation of the connectivity of time-varying networks [14], and environmental modeling by mobile sensor networks [8].

Until now, the study of estimation and control based on average consensus estimators has been through theory and simulation only. In this paper, we present the first experimental implementation of the approach. Our application is the formation control problem of [6, 16], illustrated in Figure 1. Following [3], we describe the desired formation of the swarm by a set of inertial moments, where the first moments describe the swarm's center of mass, the second moments describe the swarm's inertia, and third and higher moments further constrain the distribution of robots. The low-order moments form a convenient low-dimensional specification of the swarm shape. For example, the swarm's human supervisor could simply command the swarm to move its center of mass to a specific location, and

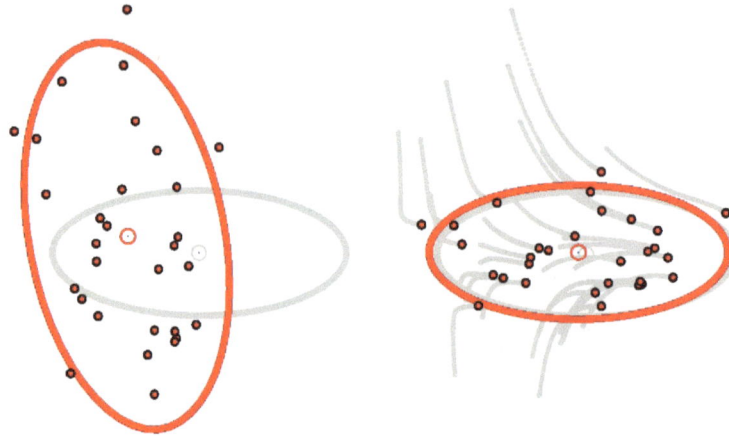

FIGURE 1. (Left) The initial configuration of a swarm, a red uniform-density ellipse with the same mass and first- and second-order moments as the swarm, and the goal formation of the swarm represented as a grey uniform-density ellipse. (Right) The swarm converges to a configuration having the desired first- and second-order moments.

to spread out according to some desired inertia. This frees the supervisor from having to specify a desired location for each robot.

We consider formations described by first and second moments. As shown in Figure 1, this abstraction of a planar formation can be visualized as an ellipse of uniform density with the same first moments, second moments, and mass as the swarm. By communicating with its neighbors, each robot estimates the current swarm moments and simultaneously controls its own motion to drive the estimated moments closer to the desired moments (Figure 1).

The main result of this paper is to show that the theoretical continuous-time estimation and control algorithms of [**6**, **16**] can be successfully adapted to an experimental implementation of formation control. While the theoretical performance guarantees do not carry over directly to the experiment, in practice we have observed reliable convergence of our eight mobile robots to the goal formation shapes. Our experiments also suggest a number of avenues for future theoretical work.

In Section 2, we summarize the theoretical approach from [**6**, **16**] and describe the adaptation of the communication, estimation, and control laws to the experimental implementation. In Section 3, we describe the "e-puck" mobile robots used in our experiments, the wireless communication system, the vision tracking system, and details of the robot control software. Section 4 presents the results of our experiments. Finally, Section 5 summarizes the results and suggests future directions.

2. Review of Simultaneous Estimation and Control of Formation Moments

2.1. Theory. Each robot is modeled as a point mass with position given by the column vector $p_i \in \mathbb{R}^m$. For our planar experiments, $m = 2$ and $p_i = [p_{i,x}, \ p_{i,y}]^T$. The state of the robot is $x_i = [p_i^T, \ \dot{p}_i^T]^T$, with dynamics

$$(1) \qquad \dot{x}_i = \left[\begin{array}{c} \dot{p}_i \\ u_i \end{array} \right],$$

where $u_i \in \mathbb{R}^2$ is the control force. For n robots, the configuration of the entire swarm is written $p = [p_1^T, \ \dots, \ p_n^T]^T \in \mathbb{R}^{2n}$.

We define a goal formation for the swarm by a set of desired first- and second-order inertial moments. The first moments specify the swarm's center of mass and the second moments specify its direction-dependent spread in the plane. For a set of unit-mass point robots in the plane, the five mass-normalized first and second moments expressed in a fixed global frame can be written as the vector

$$(2) \qquad f(p) = \frac{1}{n} \sum_{i=1}^n \phi(p_i) = \frac{1}{n} \sum_{i=1}^n [p_{i,x}, \ p_{i,y}, \ p_{i,x}^2, \ p_{i,x}p_{i,y}, \ p_{i,y}^2]^T,$$

where $\phi(p)$ is called the *moment-generating function*. The desired formation of the group, $f^\star$, specifies a $(2n - 5)$-dimensional submanifold of the $2n$-dimensional configuration space of the robots.

Each robot measures its own position and knows $f^\star$. To drive the robots to the goal configuration submanifold, we define a potential function

$$(3) \qquad J(p) = [f(p) - f^\star]^T \Gamma [f(p) - f^\star]$$

with $\Gamma \in \mathbb{R}^{5 \times 5}$ a suitably chosen symmetric positive-definite gain matrix. The control law implemented by robot i simply drives it along the negative gradient of this potential, with added damping so that the robots come to rest on the goal submanifold. The control law is

$$(4) \qquad u_i = -B\dot{p}_i - \left[\mathscr{J}\phi(p_i) \right]^T \Gamma \left[f(p) - f^\star \right],$$

where $B \in \mathbb{R}^{2 \times 2}$ is a positive-definite damping matrix and $\mathscr{J}\phi(\cdot)$ denotes the 5×2 Jacobian matrix of ϕ with respect to p_i.

Note that the control law (4) involves the swarm's current moments $f(p)$, a global quantity. In a decentralized implementation, each robot must replace $f(p)$ with a local estimate obtained through communication with neighboring robots. For scalability, this local communication cannot simply relay all accumulated sensory data, but rather must summarize the data so that the size of the communication packets and the data each robot must store is independent of the number of robots in the swarm.

To obtain local estimates of the global swarm moments, each robot implements a *proportional-integral (PI) average consensus estimator*. Each robot continuously receives its neighbors' estimates of the swarm's moments, "averages" them with its own, and broadcasts its new estimate. Inputs to the averaging process are the robots' measurements of their own positions. More specifically, robot i broadcasts to its neighbors its estimator output v_i and its estimator internal state w_i, and calculates

$$(5) \qquad \dot{v}_i = \gamma(\phi(p_i) - v_i) - \sum_{j \neq i} a(p_i, p_j) \left[v_i - v_j \right] + \sum_{j \neq i} b(p_i, p_j) \left[w_i - w_j \right]$$

$$(6) \qquad \dot{w}_i = - \sum_{j \neq i} b(p_i, p_j) \left[v_i - v_j \right],$$

where $a(p_i, p_j) \geq 0$ is a proportional gain between robots i and j, $b(p_i, p_j) \geq 0$ is an integral gain between robots i and j, and $\gamma > 0$ is a global "forgetting factor." The gains $a(\cdot, \cdot)$ and $b(\cdot, \cdot)$ implement an undirected graph. When robots i and j cannot communicate, $a(p_i, p_j) = 0$ and $b(p_i, p_j) = 0$. New information enters the averaging process through the

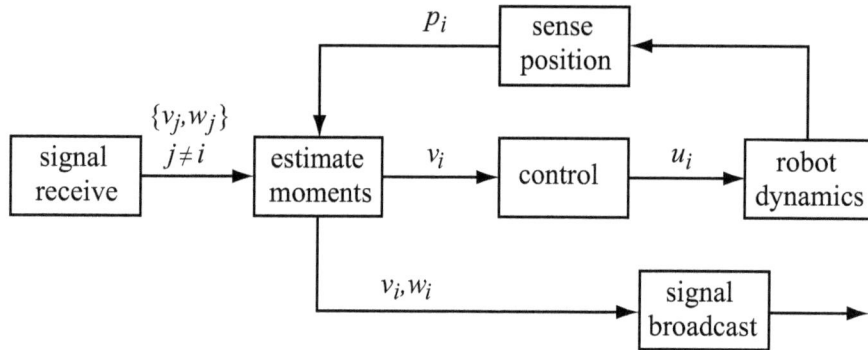

FIGURE 2. Block diagram for robot i.

term $\gamma\phi(p_i)$ in equation (5), with γ controlling the rate at which new information replaces old information.

In [6, 16], we demonstrated that the PI average consensus estimator in (5) and (6) results in estimates v_i that track $f(p)$ closely, provided that the communication network is connected and $f(p)$ is changing slowly. Furthermore, if the inputs stop changing (p is constant), the estimates converge to $f(p)$ with zero steady-state error. This behavior occurs even after dropped packets, noisy messages, and robots entering or leaving the network.

Substituting the estimator output v_i for the actual moments $f(p)$ in (4) yields the decentralized control law for robot i

$$(7) \qquad u_i = -B\dot{p}_i - \left[\mathscr{J}\phi(p_i)\right]^T \Gamma\left[v_i - f^\star\right].$$

As demonstrated in [16], a slight modification of this control law, used in concert with the PI estimator, will cause the swarm to converge to the desired configuration submanifold from nearly every initial state, and all equilibria away from the goal submanifold are unstable.

A block diagram of robot i's control system is given in Figure 2.

2.2. Modifications for Experiments. To implement the theoretical estimation and control procedure described above on actual robots, we made three primary modifications:

(1) The control law (7) was modified for kinematic robots with velocities as controls, instead of point masses with forces as controls.

(2) The definition of second moments in (2), referenced to the origin, was modified to central second moments, referenced to the swarm's center of mass.

(3) The continuous-time algorithms were adapted to the experimental realities of limited communication bandwidth and digital computing, such as dropped packets and asynchronous discrete-time sensing and control.

The first two modifications are easily handled theoretically and have little impact on performance guarantees. Understanding the effects of the third modification defines an entire research agenda which we are only beginning to explore. The three modifications are discussed below.

Kinematic robots. The mobile robots, discussed in Section 3, are most naturally modeled as kinematic robots with velocities as controls. Therefore the state of the system is simply the robot's configuration, $x_i = p_i$, and the dynamics are written $\dot{x}_i = u_i$. With these dynamics, we simplify the gradient control law (7) to

$$u'_i = -\left[\mathscr{J}\,\phi(p_i)\right]^T\Gamma\left[v_i - f^\star\right].$$

(8)

As shown in [**13**], kinematic robots under the control law (8) enjoy performance guarantees similar to those of point mass robots with force inputs under the control law (7).

In practice, the velocities calculated by (8) are passed through a collision-avoidance filter to prevent robot collisions, and then through a saturation stage that limits the commanded robot speed. Collision avoidance is described in more detail in Section 3.4.2.

Central moments. In our previous formula (2) for calculating the moments, the moment-generating function $\phi(p_i)$ was defined as

$$\phi(p_i) = [p_{i,x},\ p_{i,y},\ p_{i,x}^2,\ p_{i,x}p_{i,y},\ p_{i,y}^2]^T.$$

This choice has the disadvantage that as the swarm translates without changing shape, the second moments change. These global second moments are with respect to the origin, not the center of mass of the swarm. The result is that the behavior of the system is not translation-invariant—a set of control gains Γ that yields good behavior for one goal center of mass may yield slow convergence or undesirable oscillation for another goal center of mass. To avoid this problem, we redefine robot i's moment-generating function to be

$$\phi(p_i) = [p_{i,x},\ p_{i,y},\ (p_{i,x}-v_{i,x})^2,\ (p_{i,x}-v_{i,x})(p_{i,y}-v_{i,y}),\ (p_{i,y}-v_{i,y})^2]^T,$$

where $(v_{i,x}, v_{i,y})$ is the robot's estimate of the group's center of mass. With this choice of $\phi(p_i)$ in (6), the robots estimate the swarm's second moments with respect to its center of mass (central second moments) rather than with respect to the origin.

Asynchronous discrete-time implementation. Each robot updates its velocity according to the control law (8) every T_1 seconds, and every T_2 seconds it updates its estimate of swarm moments and broadcasts these estimates. The discrete-time estimator is written

$$v_i = v_i^- + \gamma(\phi(p_i) - v_i^-) - K_P \sum_{j \in \mathcal{N}(i)} \left(v_i^- - v_j\right) + K_I \sum_{j \in \mathcal{N}(i)} \left(w_i^- - w_j\right)$$

(9)

$$w_i = w_i^- - K_I \sum_{j \in \mathcal{N}(i)} \left(v_i^- - v_j\right),$$

(10)

where v_i is the new swarm moments estimate, w_i is the new estimator internal state, (v_i^-, w_i^-) is the previous estimator state, p_i is the current robot position, (v_j, w_j) is the most recently received estimator state from robot j, $\mathcal{N}(i)$ is the set of robots that robot i has heard from since the last estimator update, and $K_P, K_I, \gamma > 0$ are estimator gains. Because there is no global synchronization of the robots' clocks, each robot runs its control, estimation, and communication processes on a different schedule. After updating its estimate, the robot discards estimator state information received from other robots, so that each estimator update depends only on communication received during the past T_2 seconds.

In our implementation, $T_1 = 0.2$ s and $T_2 = 0.4$ s.

3. Details of the Experimental Implementation

Our experimental implementation consists of eight small mobile robots moving on a 300 cm × 300 cm section of a smooth, white floor. To obtain position estimates of each robot, four cameras view the workspace from above. The top of each robot has a unique black-and-white pattern, allowing measurement of the position and orientation of each robot. Each robot is equipped with an XBee radio [**4**], and the vision system

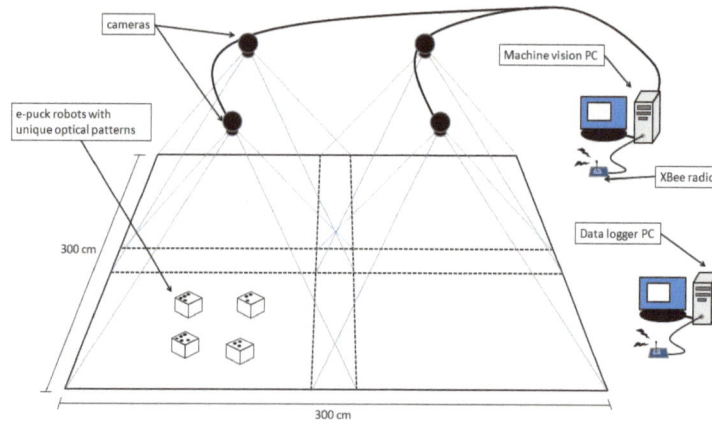

FIGURE 3. Schematic of the experimental setup.

periodically broadcasts position updates to each robot's XBee. Thus the vision system and wireless broadcasts act as a GPS. Robots broadcast state estimates to each other through the same XBee radios. A data logger PC listens to the messages being broadcast by the robots to monitor the behavior of the system. The robots' estimation and control is truly decentralized—each robot runs the same software, and the communication, computation, and storage requirements are independent of the number of robots in the swarm.

An overview of the system is shown in Figures 3 and 4. More details about the mobile robot, wireless communication, indoor positioning system, robot firmware, and command system are given below.

3.1. Mobile Robot. For our experiments, we selected e-puck mobile robots, developed at the École Polytechnique Federale de Lausanne (Figure 5) [**9**]. We chose these robots because of their simple drive system, small size (70 mm diameter), sufficient processing power, extensibility, and relatively low cost. The two wheels of the differential-drive robot are actuated by stepper motors, allowing for accurate velocity control and dead reckoning without the need for encoders. A 16-bit Microchip dsPIC30F6014A controls the robot's motors and peripherals, and has enough processing power for our control, estimation, and communication needs.

The e-puck comes standard with Bluetooth wireless communication, which requires a "master" device to coordinate several "slave" devices. This architecture is unsuitable for decentralized estimation and control. We therefore added a Digi XBee RF module [**4**] for inter-robot communication. The e-puck's PIC microcontroller communicates with the XBee using RS-232. The wireless communication system is discussed further in Section 3.2.

The chassis of each robot has three degrees of freedom in the plane, represented by a point (x, y) halfway between the wheels and an angle θ corresponding to the "forward" motion direction. To recover the point robot model, we control the motion of a reference point $(x_\mathrm{r}, y_\mathrm{r})$ fixed in the robot's frame, defined as

$$(x_\mathrm{r}, y_\mathrm{r}) = (x + h \cos \theta, y + h \sin \theta), \quad h > 0,$$

FIGURE 4. A photograph of the experimental arena.

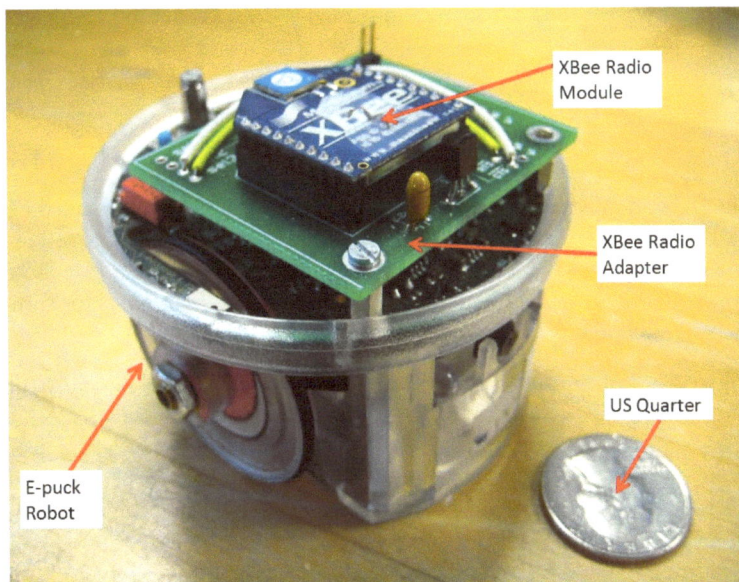

FIGURE 5. The EPFL e-puck mobile robot, with its original extension module replaced with the custom XBee radio extension module.

and leave θ uncontrolled (Figure 6). For wheels of radius R separated by a distance L along the common axle line, the relationship between the commanded velocity $[u_x, u_y]^T$ of the reference point and the commanded angular velocities u_ℓ and u_r of the left and right

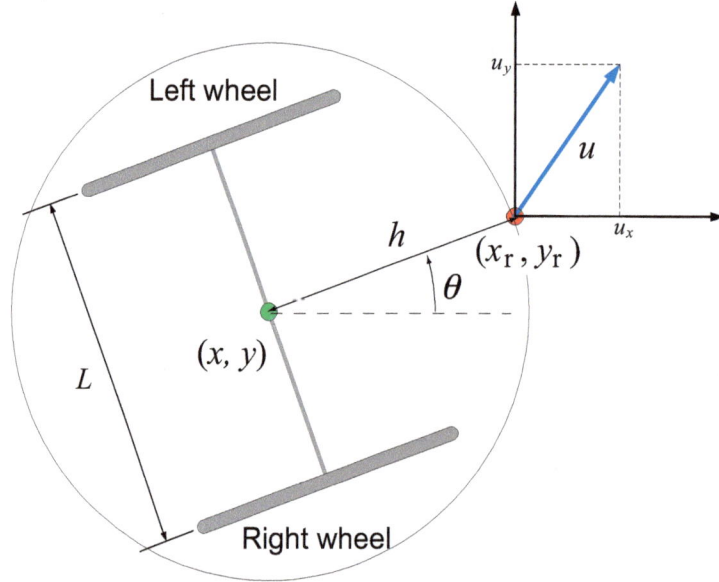

FIGURE 6. The reference point (x_r, y_r) is commanded to move with a velocity $u = [u_x, u_y]^T$, which is converted to wheel speeds in (11).

wheels, where $u_\ell, u_r > 0$ causes the robot to go forward, is

$$(11) \qquad \begin{bmatrix} u_\ell \\ u_r \end{bmatrix} = \frac{1}{2hR} \begin{bmatrix} 2h\cos\theta + L\sin\theta & 2h\sin\theta - L\cos\theta \\ 2h\cos\theta - L\sin\theta & 2h\sin\theta + L\cos\theta \end{bmatrix} \begin{bmatrix} u_x \\ u_y \end{bmatrix}.$$

By choosing $h = L/2$, the velocity limits on the wheel speeds, represented as a square in the (u_ℓ, u_r) space, become a square in the (u_x, u_y) space of velocities of the reference point. For the e-puck, $L = 53$ mm and $R = 20.5$ mm, and we use $h = 35$ mm. The maximum wheel speed for the e-puck is 1 rev/s, yielding a maximum forward speed of approximately 130 mm/s.

3.2. Wireless Communication. As mentioned earlier, the robots are outfitted with XBee radios, which they use for broadcasting and receiving messages. A data logger PC also uses an XBee radio to capture the broadcast traffic. The data logger attaches timestamps to each received message so that the behavior of the system can be analyzed and plotted later. Finally, the vision PC, which is connected to the four cameras, uses an XBee radio to broadcast the positions and orientations of each of the robots.

XBee radios are low-cost, low-power (1mW) radios that use the IEEE 802.15.4 standard and operate in the 2.4 GHz ISM band. Each XBee module contains an RF transceiver and a microcontroller implementing basic networking capabilities such as packet addressing and checksums. Together, these radios form a peer-to-peer network in which each member can broadcast a message to all other members. Each XBee embeds a unique identification number in the packets it sends, allowing packet recipients to identify packet senders.

The XBee connects to the robot's microcontroller via the e-puck's extension connector. The e-puck's microcontroller communicates with the XBee module over a serial port using the RS-232 protocol at 115200 baud. Received data can be stored in a 100 byte buffer on the radio, so the microcontroller need not process received data immediately. Each packet is approximately 40 bytes long, however, so the microprocessor must process XBee data regularly to prevent buffer overflows.

The practical range of the XBee radios is on the order of tens of meters, so each robot in the test arena can hear from every other robot. To simulate radius-limited communication, each robot broadcasts its position along with its estimator state. Receiving robots simply ignore messages received from robots beyond a certain distance.

In our implementation, each robot broadcasts its estimator state every $T_2 = 0.4$ s on a randomly initialized clock. We did not implement any protocols to adapt broadcast frequency or timing based on network traffic. The vision PC also periodically broadcasts the position and orientation of each robot. Based on results from the data logger PC, we estimate that each XBee radio drops 5-10 percent of the packets that are broadcast. These dropped packets mean that the communication network may temporarily become directed (one-way communication between robots) rather than undirected (if robot i hears from robot j, then robot j hears from robot i), as assumed in the theory. This manifests itself as noise in the consensus algorithms.

3.3. Indoor Positioning System. To simulate GPS for a small-scale indoor environment, we developed a positioning system using webcams and OpenCV, an open-source computer vision library. Our system allows us to track the positions of multiple robots with less than 1 cm error.

The hardware for the positioning system consists of four Logitech Quickcam Communicate Deluxe USB web cameras mounted on the corners of a rectangular frame suspended above the test arena. Using multiple cameras increases the coverage of the system without the need for a single high-resolution camera and wide-angle lens. The cameras have overlapping fields of view in the interior of the arena to ensure that at least one camera images every target as it moves between fields of view.

To track the robots' positions and orientations, each robot has a unique rotationally-asymmetric pattern of black dots on a white background mounted on its top side (see Figure 7). To calibrate the cameras, a grid of 25 point markers is evenly spaced over the test arena, with each marker at the height of the robots' dot patterns. The grid is chosen so that each camera sees nine of the markers. A matrix transforming the camera image coordinate system to the real-world coordinate system is computed using the direct linear transformation given in [1].

The vision tracking process starts with a 640×480 pixel image from each camera. In the first step, each image is thresholded to obtain a binary image. Then the black dots are partitioned into clusters, and each robot is uniquely identified by comparing a vector of rotationally-invariant measures to known profiles associated with each robot: the number of dots in the cluster and the vector of distances from the dots to the center of mass of the dots. Once the identity of each robot is determined, its position and orientation are calculated using the positions of the dots relative to the center of mass.

All four images are collected and processed into identifiers and positions and orientations for each robot approximately 5-10 times per second. This data is displayed in real time on the vision PC monitor. The vision PC broadcasts all of the robots' positions and orientations approximately once every three seconds. In between virtual GPS updates, the robots use dead reckoning to update their position estimates. The relatively low virtual GPS update rate was chosen to reduce network traffic, and because dead reckoning is accurate over short durations.

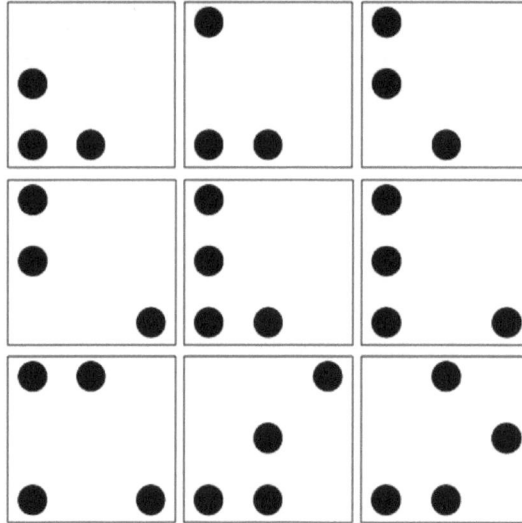

FIGURE 7. A sample of the patterns used with the vision algorithm.

3.4. Robot Firmware. The firmware code for the PIC microcontroller on the e-puck implements the distributed consensus estimator and the motion controller as described in Section 2.2. The code is written in C and compiled with Microchip's C30 compiler. Each robot has identical firmware, but can be differentiated by the unique identifier of its XBee radio.

When implementing these algorithms, practical limitations of the hardware, such as the microcontroller's relatively slow floating point operation performance and limited memory, must be considered. Also, the low-power XBee radio network has limited data throughput; if too many robots attempt to transmit data simultaneously, many data packets will be lost.

To account for these limitations, we limit the rate of calculating control and estimation laws and transmitting data. Every 200 ms, a timer interrupt generates a clock "tick." Every tick, the motion controller evaluates the control law (8) and updates the wheel speeds. Every second tick, the consensus estimator is evaluated and the results broadcast. This timing was chosen experimentally to yield good performance while giving the microcontroller ample time to complete its tasks. A summary of the control flow is shown in Figure 8.

Further firmware details are given below.

3.4.1. *Initializing the Consensus Estimator.* When a robot is turned on and first joins the communication network, its estimator state variables are initialized to zero. Any choice would suffice, as initial estimator values quickly decay.

3.4.2. *Collision Avoidance.* The reference point velocity u' produced by the control law (8) is passed through a simple collision-avoidance filter to produce a modified control u that prevents collisions between robots. Based on the position information broadcast by other robots, each robot determines the position of its nearest neighbor. Let $d \in \mathbb{R}^2$ be the vector from the robot to its nearest neighbor, and define $\eta = -\|u'\|d/\|d\|$ to be a "collision-avoidance velocity" in the direction of $-d$ with magnitude $\|u'\|$. We define three zones around each robot, represented by an "ignore radius" R_2 and a "proximity radius"

Program Main Loop

Serial Port Receive ISR (high priority)

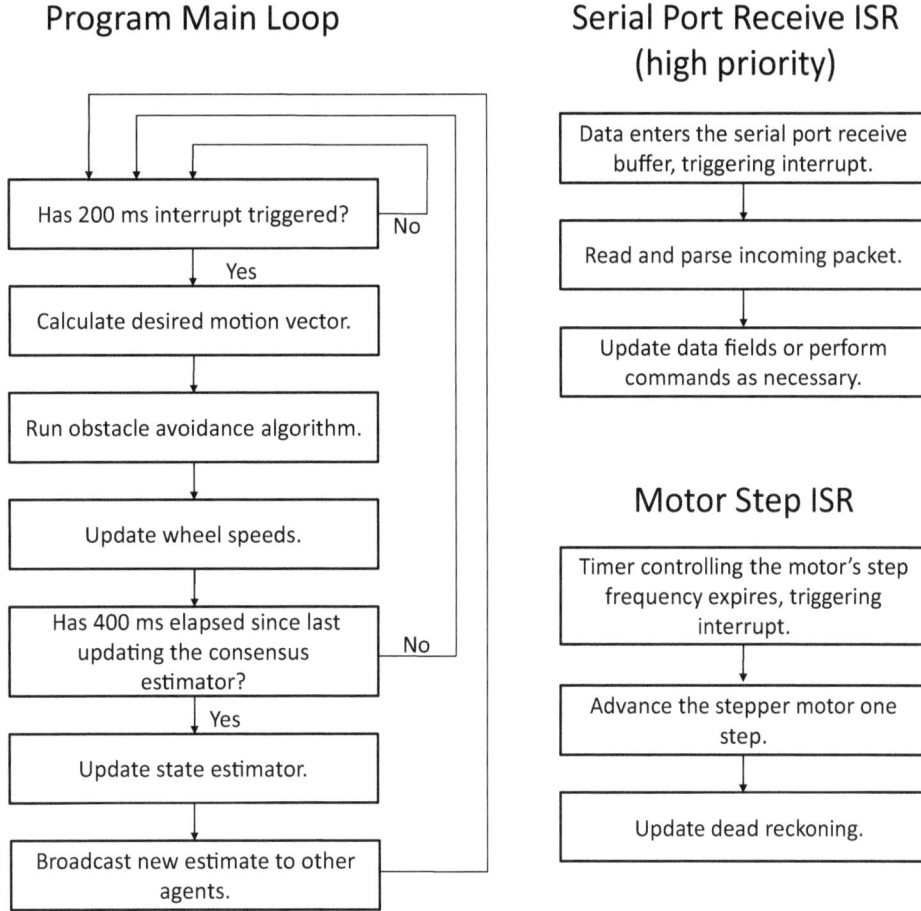

FIGURE 8. State diagram for the firmware on the e-puck robot. The high priority interrupt service routine triggered by the UART receive buffer interrupts the other interrupt service routines.

R_1, where $R_2 > R_1$. If $\|d\| > R_2$, then there is no danger of collision and $u = u'$. If $\|d\| \leq R_1$, then the robot follows the collision-avoidance velocity ($u = \eta$). At intermediate distances, the velocity u is a convex combination of the originally calculated velocity u' and the collision-avoidance velocity η, according to the formula

$$(12) \qquad \alpha = \frac{\|d\| - R_1}{R_2 - R_1}$$

$$(13) \qquad u = \begin{cases} u' & R_2 < \|d\| \\ \alpha u' + (1 - \alpha)\eta & R_1 < \|d\| \leq R_2 \\ \eta & \|d\| \leq R_1. \end{cases}$$

3.4.3. *Wheel Speed Controller and Saturation.* Every $T_1 = 0.2$ s, the robot's commanded velocity u is calculated from the control law (8) and passed through the collision-avoidance filter (13). This velocity is then converted to commanded wheel speeds using (11). If either wheel speed exceeds the maximum of 1 rev/s, both wheel speeds are scaled to maintain their ratio while satisfying the maximum speed constraint. For typical gain values Γ in our experiments, a wheel speed is saturated until the swarm is close to satisfying the desired formation moments.

While the wheel speeds are constant between evaluations of the control law, the linear velocity of the reference point is generally not constant, due to the changing angle of the robot. This deviation from the theory does not cause significant practical issues.

The microcontroller implements a commanded wheel speed by converting the speed to a time between wheel increments and setting a timer to interrupt with this period. The timer's interrupt service routine increments the wheel angle and updates the dead reckoning estimate.

3.5. Command System. The vision PC doubles as a command console for the swarm, allowing the broadcast of control parameter updates, changes in goal behavior, and other user interaction. Example commands include

- sleep (put one or more robots into a inactive state)
- wake (put one or more robots into an active state)
- change goal state ($f^\star$)
- change consensus estimator gains (K_P, K_I, γ)
- change motion controller gains (Γ)
- change maximum communication radius

4. Experimental Results

We tested the swarm under several scenarios:

(1) Moving to a goal formation.
(2) Repeating the first experiment using poorly chosen motion control gains.
(3) Repeating the first experiment when the robots have a limited communication radius.
(4) Moving the swarm through a bottleneck by giving successive goal formations.

Each experiment used eight e-puck robots. As the experiments ran, the data logging PC recorded all of the packets broadcast. The plots in this section use data extracted from these packets. In several of the following figures, data logged from the experiment is plotted over a frame from a video recording of the experiment, taken by a dedicated video camera (not one of the vision system webcams). The green ellipse is a graphical representation of the swarm's actual moments, and the red ellipse represents the goal moments. The moment estimates of each robot are also plotted against time, along with the goal moments (blue dashed line) and the actual moments of the swarm (red dashed line).

In these experiments, the major source of noise was packet loss. When a packet from a robot is lost, it appears as if the robot were momentarily removed from the swarm, so the rest of the swarm compensates by shifting position, causing jittering when the swarm is near its goal state. This jittering is reduced, at the expense of some steady-state error, by imposing a deadband on the control u: if the norm of u is less than some threshold, the robot remains stationary.

We chose controller and estimator gains experimentally by observing their effect on the swarm. Unless otherwise specified, we used the following parameters, where $(0, 0)$ is the center of the arena:

- Goal: $f^\star = [100 \text{ mm}, 300 \text{ mm}, 160000 \text{ mm}^2, 40000 \text{ mm}^2, 40000 \text{ mm}^2]^T$. In the plots of the experimental results, first and second moments are referred to as $[x_\text{cm}, y_\text{cm}, I_{yy}, I_{xy}, I_{xx}]^T$.

- Motion control gains: $\Gamma = \mathrm{diag}(1\ \mathrm{s}^{-1},\ 1\ \mathrm{s}^{-1},\ 1 \times 10^{-5}\ \mathrm{mm}^{-2}\mathrm{s}^{-1},$
$$2 \times 10^{-5}\ \mathrm{mm}^{-2}\mathrm{s}^{-1},\ 1 \times 10^{-5}\ \mathrm{mm}^{-2}\mathrm{s}^{-1}).$$
- Collision avoidance parameters: $R_2 = 200$ mm, $R_1 = 100$ mm.
- Communication radius: infinite.
- Estimator gains for robot i: $K_P = 0.7/|\mathcal{N}(i)|$, $K_I = 0.1/|\mathcal{N}(i)|$, $\gamma = 0.05$, where $|\mathcal{N}(i)|$ is the number of robots that robot i has received information from in the past T_2 seconds. (For a known bound on the number of robots, it may be better to choose K_P and K_I as constants, to avoid the possibility of introducing unbalanced networks when $|\mathcal{N}(i)|$ is different for each robot; see [5].)

4.1. Simple Formation Control. In this experiment, the robots are scattered throughout the test arena and the goal is fixed. Figure 9 shows the starting positions of the robots and the trace of their paths as they converge to the desired configuration manifold. Figure 10 plots the first and second moment estimates of each of the robots, as well as the actual moments, as a function of time.

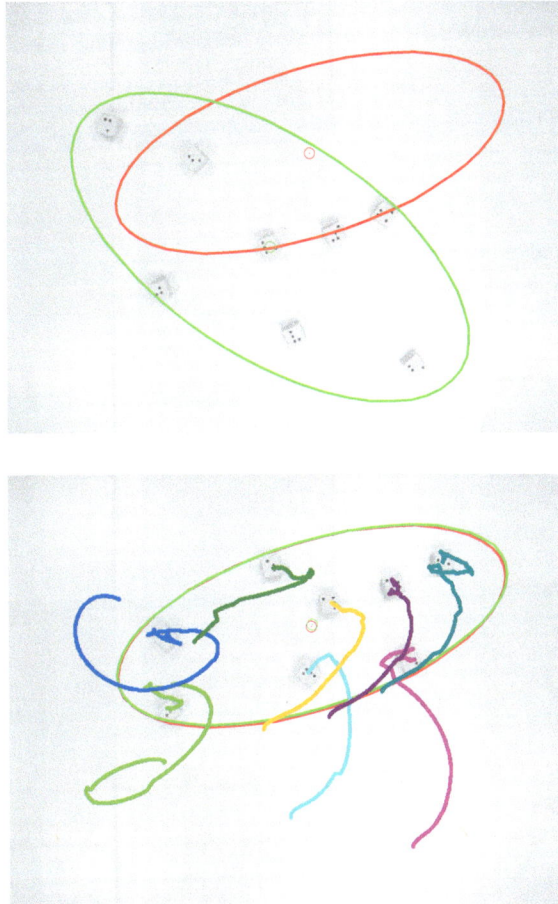

FIGURE 9. (Top) The initial and desired configurations of the swarm. (Bottom) A trace of the motion of the robots.

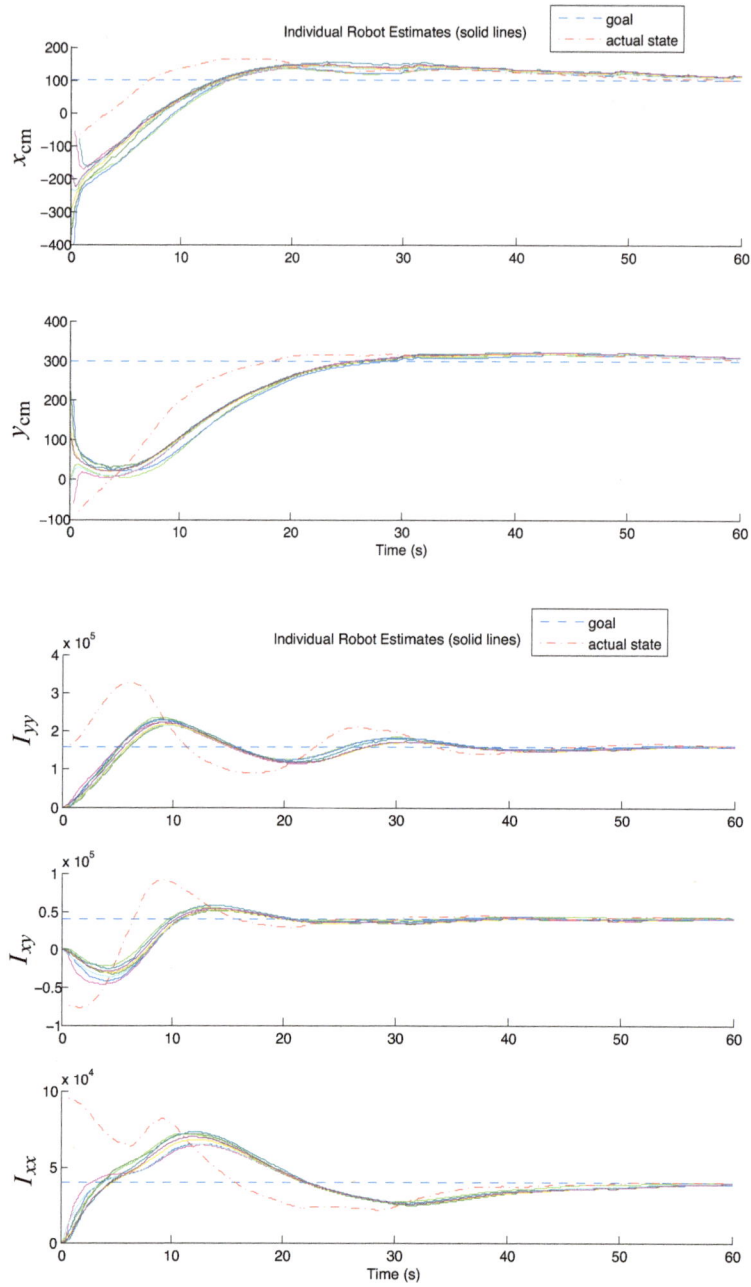

FIGURE 10. Time histories of the goal swarm moments, the actual swarm moments, and each robot's estimate of the swarm moments. (Top) First moments. (Bottom) Second moments.

4.2. Effect of Poorly Selected Gains. To explore the effect of controller gains, we increased the controller gains Γ by a factor of 10. Because robot speeds are usually saturated anyway, the primary effect of this uniform increase in gains is increased oscillation when the robots get close to the goal state. Figures 11 and 12 show the experimental behavior.

Increasing the estimator gains eventually causes the discrete-time estimators to become unstable, leading to unstable swarm motion.

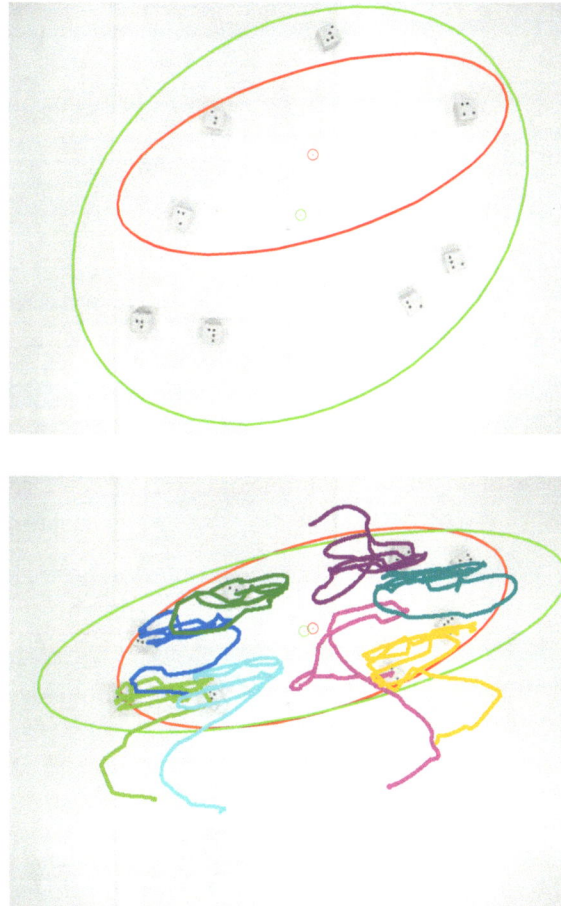

FIGURE 11. (Top) The initial and desired configurations of the swarm. (Bottom) A trace of the motion of the robots with large controller gains.

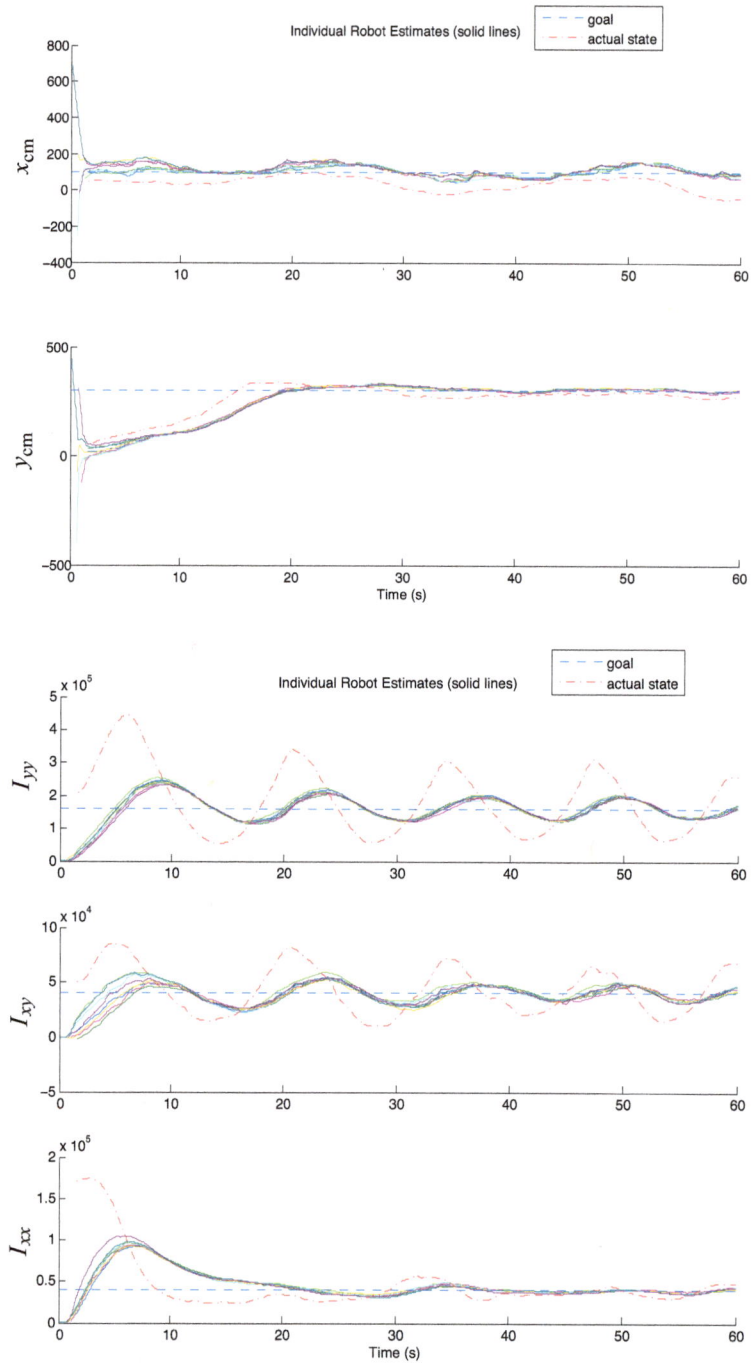

FIGURE 12. Time histories of the goal swarm moments, the actual swarm moments, and each robot's estimate of the swarm moments with large controller gains. (Top) First moments. (Bottom) Second moments.

4.3. Limited Communication Radius: 750 mm. This experiment uses the same conditions as the first, except that the robots are given a maximum communication radius of 750 mm. Every robot discards packets received from robots outside of this radius. The decreased connectivity and transients induced by the breaking and establishment of communication links leads to increased noise in the estimates and a slower settling time. The performance is shown in Figures 13 and 14.

FIGURE 13. (Top) The initial and desired configurations of the swarm. (Bottom) A trace of the motion of the robots with communication radius set to 750 mm.

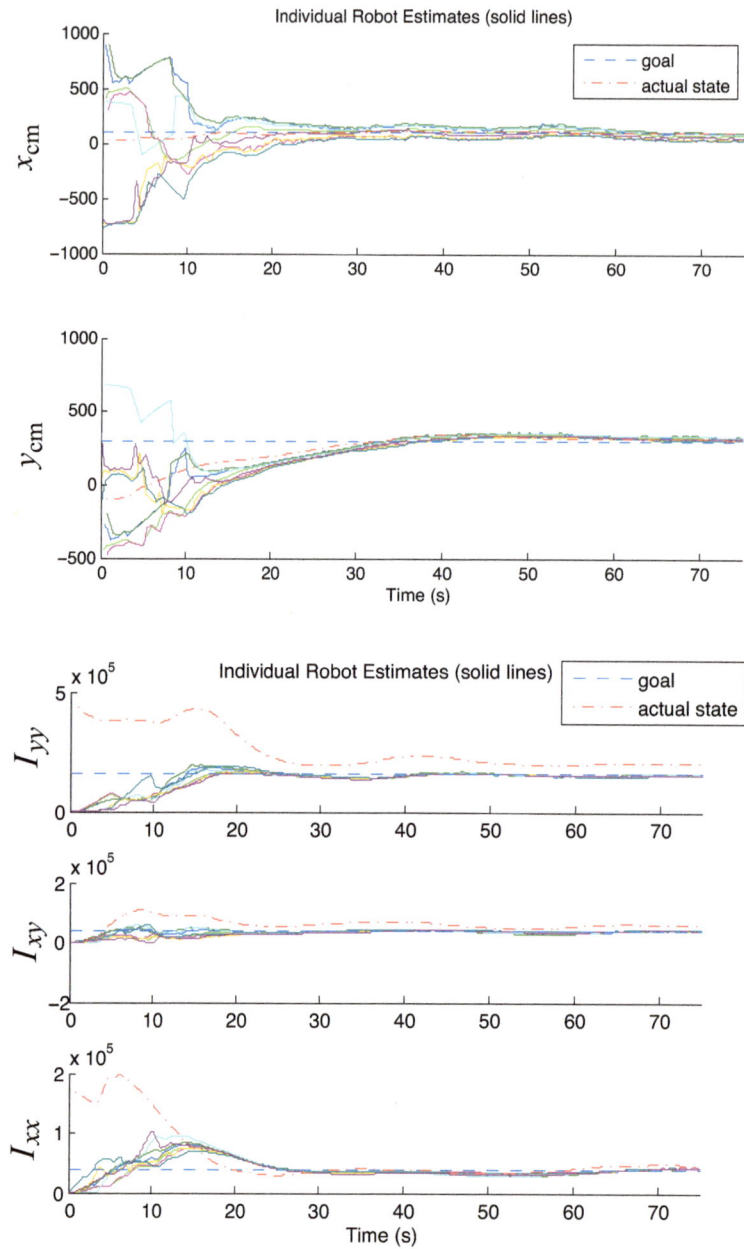

FIGURE 14. Time histories of the goal swarm moments, the actual swarm moments, and each robot's estimate of the swarm moments with communication radius set to 750 mm. (Top) First moments. (Bottom) Second moments.

4.4. Limited Communication Radius: 500 mm. When the communication radius is decreased to 500 mm, the performance of the swarm degrades further. The continual formation and disintegration of communication links creates transients that prevent the swarm from settling smoothly. Figures 15 and 16 show the results of this experiment.

FIGURE 15. (Top) The initial and desired configurations of the swarm. (Bottom) A trace of the motion of the robots with communication radius set to 500 mm.

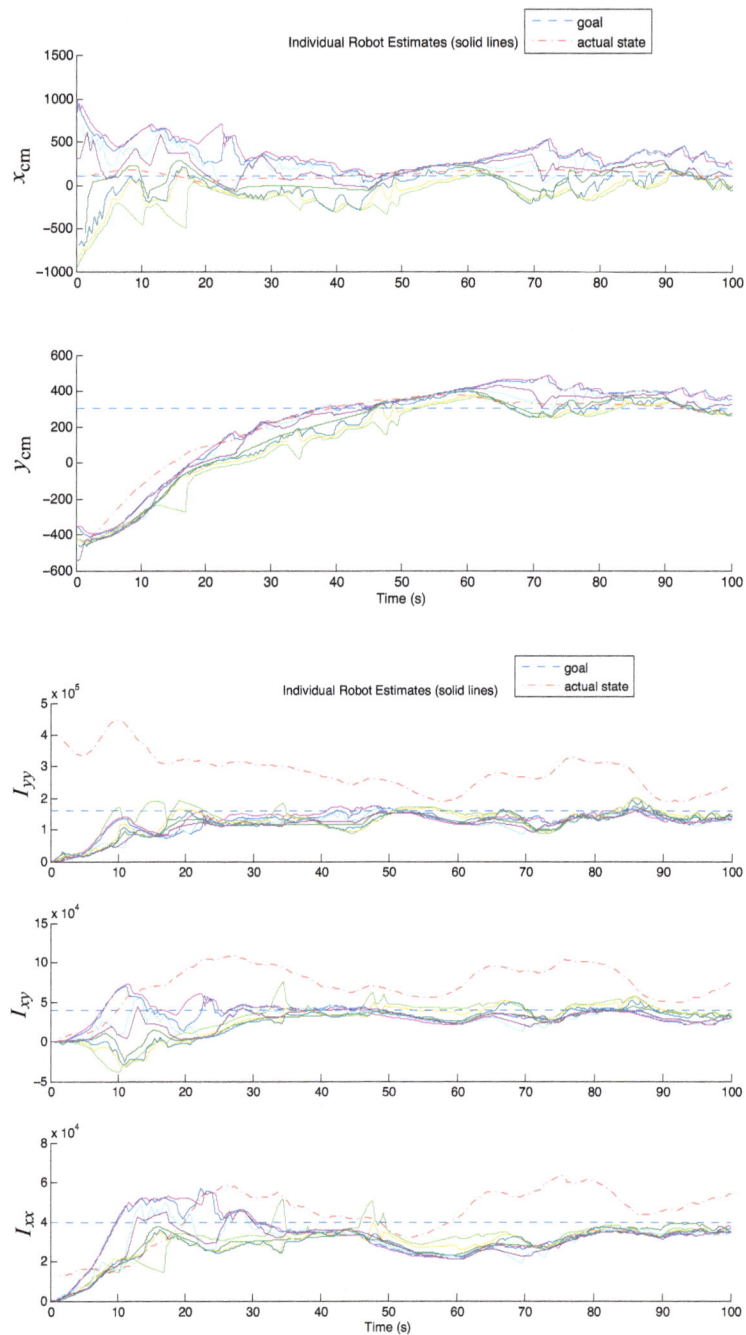

FIGURE 16. Time histories of the goal swarm moments, the actual swarm moments, and each robot's estimate of the swarm moments with communication radius set to 500 mm. (Top) First moments. (Bottom) Second moments.

4.5. Morphing Formation. In the next experiment, a sequence of goal states is sent to the swarm. The result is to move the swarm through a bottleneck, as marked by black virtual obstacles in Figure 17. We used the following sequence of formation moments:

(1) $f_1^\star = [-800 \text{ mm}, \ 0 \text{ mm}, \ 30000 \text{ mm}^2, \ 0 \text{ mm}^2, \ 150000 \text{ mm}^2]^T$
(2) $f_2^\star = [-600 \text{ mm}, \ 0 \text{ mm}, \ 50000 \text{ mm}^2, \ 0 \text{ mm}^2, \ 50000 \text{ mm}^2]^T$
(3) $f_3^\star = [0 \text{ mm}, \ 0 \text{ mm}, \ 160000 \text{ mm}^2, \ 0 \text{ mm}^2, \ 20000 \text{ mm}^2]^T$
(4) $f_4^\star = [600 \text{ mm}, \ 0 \text{ mm}, \ 50000 \text{ mm}^2, \ 0 \text{ mm}^2, \ 50000 \text{ mm}^2]^T$
(5) $f_5^\star = [800 \text{ mm}, \ 0 \text{ mm}, \ 30000 \text{ mm}^2, \ 0 \text{ mm}^2, \ 150000 \text{ mm}^2]^T$

Figure 17 shows the starting positions and the successive goals of the robots, as well as a trace of the paths of the robots. Figure 18 shows the actual and estimated moments as a function of time.

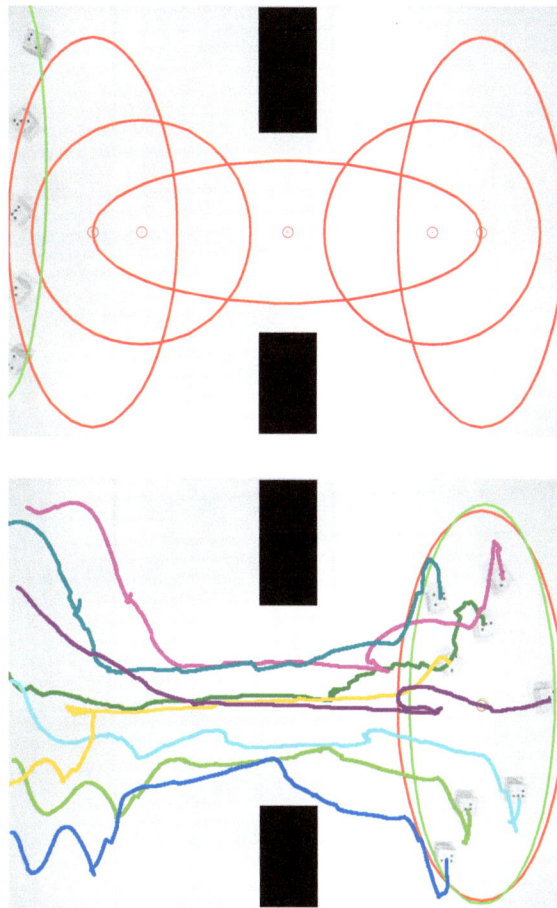

FIGURE 17. (Top) The initial and desired configurations of the swarm. (Bottom) A trace of the motion of the robots during target morphing.

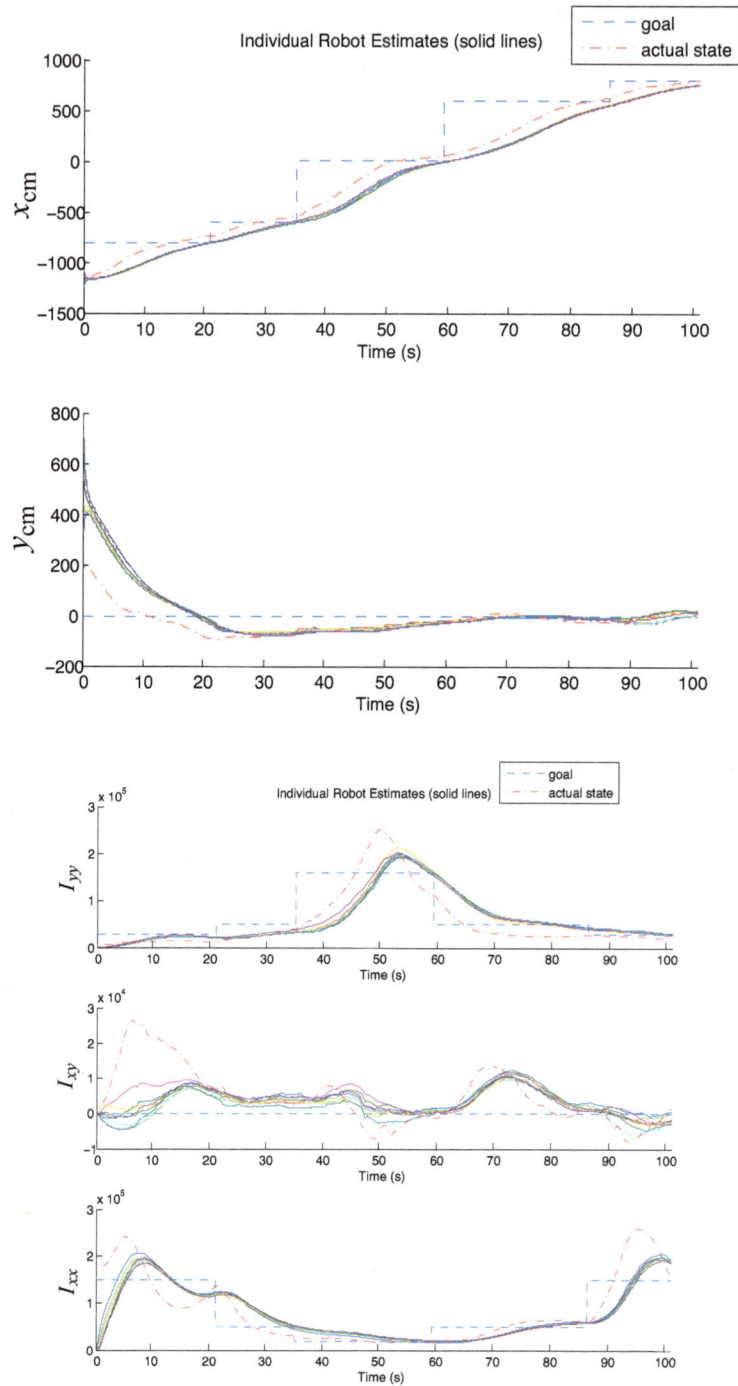

FIGURE 18. Time histories of the goal swarm moments, the actual swarm moments, and each robot's estimate of the swarm moments during target morphing. (Top) First moments. (Bottom) Second moments.

5. Conclusion and Future Work

Our experimental results demonstrate that a group of mobile robots can achieve a desired formation, as described by a set of moments, without the need for any central coordination. We achieved this result by adapting the theory presented in [**6, 16**] for implementation on real hardware. Despite this success, we had to make a number of choices to convert the theory to experiment, and many open questions remain.

In our implementation, only packets received in the past T_2 seconds are incorporated into the state estimate update. Intermittently dropped packets, or a communication link that is newly broken or established because of a communication radius limit, can therefore lead to high-frequency variations in a robot's state estimate. This is particularly noticeable in the 500 mm communication radius experiment in Figure 16. An alternative would be to keep estimates received during the past T seconds, $T > T_2$. Each time a new estimate is received from a particular robot, it replaces that robot's previous estimate. If a robot has not been heard from in the past T seconds, its estimate is deleted. Choosing $T > T_2$ would create a low-pass effect, reducing variations in each robot's estimate and making the estimates less sensitive to changes in network topology. This would come at the expense of responsiveness of the estimator—if T is chosen large, then stale information, or even information from a robot that has since been removed from the swarm, may be included in the estimator updates.

A related issue is the implementation of information-diffusion broadcasting. Our approach is extremely simple; future research could focus on more advanced network protocols that attempt to maximize network bandwidth and limit packet loss.

We chose gains for the consensus estimator empirically, based on experiments. While there exist guidelines for choosing stable estimator gains for continuous-time [**5, 7**] and synchronous discrete-time [**10**] estimators, such guidelines do not exist for asynchronous discrete-time estimators. Future work should address choices of gains that yield stability under worst-case conditions and good performance under typical conditions.

We are currently adapting our robots to allow them to sense properties of the environment. Each robot has an upward-pointing color sensor, and an overhead computer projector projects virtual dynamic environments onto the floor. Color from the projector simulates a time-varying function of position, such as temperature or chemical concentration. This new setup will allow us to perform experiments in control for optimal data collection and decentralized environmental modeling [**8**].

Acknowledgments

We thank the several Northwestern undergraduates who have led the continued development of the robot swarm experiment, including Ryan Cook, Sam Bobb, Jonathan Lee, and Neal Ehardt. This work was supported by the National Science Foundation and the Office of Naval Research.

Bibliography

[1] Y. I. Abdel-Aziz and H. M. Karara, *Direct linear transformation from comparator coordinates into object space coordinates in close-range photogrammetry.*, Proceedings of the Symposium on Close-Range Photogrammetry (1971), 1–18.

[2] He Bai, Randy A. Freeman, and Kevin M. Lynch, *Robust dynamic average consensus of time-varying inputs*, IEEE International Conference on Decision and Control, 2010.

[3] Calin Belta and Vijay Kumar, *Abstraction and control for groups of robots*, IEEE Transactions on Robotics **20** (2004), no. 5, 865–875.

[4] Digi International, *Xbee / xbee-pro rf modules*, v1.xex ed., 2009.

[5] Randy A. Freeman, Thomas R. Nelson, and Kevin M. Lynch, *A complete characterization of a class of robust linear average consensus protocols*, American Control Conference, 2010.

[6] Randy A. Freeman, Peng Yang, and Kevin M. Lynch, *Distributed estimation and control of swarm formation statistics*, American Control Conference, 2006.

[7] Randy A. Freeman, Peng Yang, and Kevin M. Lynch, *Stability and convergence properties of dynamic consensus estimators*, IEEE International Conference on Decision and Control, 2006, pp. 338–343.

[8] Kevin M. Lynch, Ira B. Schwartz, Peng Yang, and Randy A. Freeman, *Decentralized environmental modeling by mobile sensor networks*, IEEE Transactions on Robotics **24** (2008), no. 3, 710–724.

[9] Francesco Mondada, Michael Bonani, Xavier Raemy, James Pugh, Christopher Cianci, Adam Klaptocz, Stéphane Magnenat, Jean-Christophe Zufferey, Dario Floreano, and Alcherio Martinoli, *The e-puck, a robot designed for education in engineering.*, Proceedings of the 9th Conference on Autonomous Robot Systems and Competitions, 2009, pp. 59–65.

[10] Thomas R. Nelson, *Decentralized worst-case estimation and control in mobile sensor networks*, Ph.D. thesis, Northwestern University, Evanston, Illinois, June 2010.

[11] Reza Olfati-Saber and Jeff S. Shamma, *Consensus filters for sensor networks and distributed sensor fusion*, IEEE International Conference on Decision and Control, December 2005, pp. 6698–6703.

[12] Demetri P. Spanos, Reza Olfati-Saber, and Richard M. Murray, *Dynamic consensus on mobile networks*, IFAC World Congress, 2005.

[13] Peng Yang, Randy A. Freeman, Geoffrey J. Gordon, Kevin M. Lynch, Siddhartha Srinivasa, and Rahul Sukthankar, *Decentralized estimation and control of graph connectivity in mobile sensor networks*, American Control Conference, 2008.

[14] Peng Yang, Randy A. Freeman, Geoffrey J. Gordon, Kevin M. Lynch, Siddhartha S. Srinivasa, and Rahul Sukthankar, *Decentralized estimation and control of graph connectivity for mobile sensor networks*, Automatica **46** (2010), no. 2, 390–396.

[15] Peng Yang, Randy A. Freeman, and Kevin M. Lynch, *Distributed cooperative active sensing using consensus filters*, IEEE International Conference on Robotics and Automation, 2007.

[16] Peng Yang, Randy A. Freeman, and Kevin M. Lynch, *Multi-agent coordination by decentralized estimation and control*, IEEE Transactions on Automatic Control **53** (2008), no. 11, 2480–2496.

[17] Minghui Zhu and Sonia Martínez, *Discrete-time dynamic average consensus*, Automatica **46** (2010), no. 2, 322–329.

Send Orders of Reprints at reprints@benthamscience.net

CHAPTER 6

Extending Lifetime of the Network and Crucial Node by Multiple Diversity Combining

Lichuan Liu

Department of Electrical Engineering

Northern Illinois University

DeKalb, IL 60115, USA

ABSTRACT. We introduce the crucial nodes in sensor networks, whose energy consumption affects the network lifetime greatly. Due to the relative high traffic load at the crucial nodes, energy loss occurs when packet collision and/or retransmission. We propose a LMMSE collision separation technique by combining the space diversity provided by antenna array and network diversity provided by the Media Access Control (MAC) layer. In our solution, collided packets, at the collector, are not discarded but combined with network diversity on-demand to separate data packets from different users and improve the network performance. Using analysis and computer simulations, we demonstrate that the space and network diversity combining (SNDC) scheme for cross-layer and MMSE collision resolution is energy efficient.

1. Introduction

In recent years, the idea of wireless sensor networks has gathered a great deal of attention. A distributed wireless sensor network may have thousands of small sensor nodes. Each individual sensor contains both processing and communication elements and is designed in some degree to monitor the environmental events specified by the end user of the network. Information about the environment is gathered by sensors and delivered to a remote collector. Wireless sensor networks are widely known as resource-constrained networks in terms of sensor lifetime, mobility and computation capability. In order to achieve the optimal performance, the design of network architecture must take these features into consideration from the beginning.

The cross-layer design in wireless sensor networks has received much attention among researchers both in physical layer and network layer. The cross-layer design facilitates efficient and collaborative utilization of network resources in protocol based communication systems. Traditionally, in wired networks, the MAC layer is designed using a simple collision avoidance model. Most of the conventional random access protocols assume that the failure of reception is caused by collisions and the channel is noiseless. Therefore, collided packets are destroyed and retransmission must be made later. Simple random access protocols of ALOHA type 'resolve' collisions by randomizing retransmission to improve system performance [1]. Carrier sensing multiple access (CSMA) and collision detection mechanism are also employed to improve the throughput performance by collision detection in wired networks. The emphases in the random access mechanisms have been mostly on retransmission schemes that minimizing future collisions [2]. However, the collided packets are typically discarded when a collision does occur, and no information is exploited from them. In cellular data networks, data sensing multiple access (DSMA) is usually implemented. The base station detects collisions and continuously broadcasts a busy/idle signal through a control channel to all users. In a wireless network, users share a common *unreliable* wireless channel due to fading, interference and background noise. The receiver at a given node is designed to extract from data the signals of interested users. Better approaches to resolving and separating collisions improve the throughput and performance of the system. The key to successful signal separation is the transmission and/or receiving *diversity* embedded in the received data packages. Diversity can be introduced both at the transmitter and at the receiver. The transmit diversity introduces redundancy into the transmitted data package using modulation, coding, and spreading techniques. The spatial diversity is another technique for reliable signal reception and separation [3]. The use of antenna arrays provides additional degrees of freedom in multi-packet reception and separation [**4**]. The framework of multiuser detection [**5**] is the essence of many signal separation methods. The transmit diversity, receiver diversity, and multiuser detection are mainly implemented in physical layer. In addition, recent research also uses network resources to provide diversity through selective retransmission [**6**].

The goal of this chapter is to combine the diversities provided by different protocol layers for multi-packet reception in random access wireless networks. In this approach, employing antenna arrays at collector allows the separation of multiple transmitting at the same time when the number of active sensor nodes is less than the number of the antennas. However, with the increase of the active nodes, the collector can not extract the data by the space diversity only. Then the received packets that have collided are stored in memory. They are later combined with future retransmission in order to extract all the collided information.

2. WSN Architecture and Network Lifetime

2.1. WSN Architecture. In wireless sensor network, the sensor nodes are usually scattered in a sensor field as shown in Figure 2. These nodes collect data and route data to the collector, the collector communicates with the process center.

According to the communication mechanism for collecting the sensing data, there are different type of architectures of sensor networks, as shown in Figure 1. The simplest one is direct connected, where each sensor directly sends information to the remote receiver/collector independent of each other [**7**]. It is energy-inefficient and impossible in many cases due to the large number of sensor nodes and limited transmission range.

The second approach is multi-hop routing. Because of the battery capacity limitation of sensor nodes, multi-hop but short range transmission usually consumes less average power than one-hop long distance transmission for a given pair of source and destination.

The third approach is cluster-based multi-hop, where sensors form clusters with neighboring sensors. One sensor will be elected as the cluster head according to some rules. Collected information will be transmitted to the cluster head first, then relayed to the remote receiver/collector. This approach localizes the traffic and can be scalable. It may reduce the overall data transmission when local data fusion and classification techniques are used. The disadvantage of this approach is that the energy consumption at cluster heads is much more than other approaches.

In this chapter, we only consider the multi-hop routing architecture.

2.2. Network Lifetime and Crucial Nodes. In non-mission-critical application, the definition of lifetime is the cumulative active time of the network (i.e., whenever the network is active its lifetime clock is ticking, otherwise not.) In mission-critical applications, lifetime is defined as the cumulative active time of network until the first loss of coverage or quality failure [**8**]. In this chapter the network lifetime is defined as the time interval between the time that sensor network starts its operation and the time that the collector losses communication with all sensor nodes. In most cases, when all the nodes that can communicate with the collector directly expire, the sensor network is completely 'dead'. Therefore, for the multi-hop WSN, these nodes with one hop away from the collector are called *crucial nodes* because their lifetime are more important than other nodes for the network lifetime.

Assume that a sensor network consists of a total of N_{total} randomly deployed nodes. Denote the sensor node set $\mathcal{S} = \{s_1, s_2, \cdots, s_{Ntotal}\}$, $|\mathcal{S}| = N_{total}$, where $|\cdot|$ is the cardinal number of a set [**9**]. Define $\mathcal{C} = \{s_i | r_i < d_{max}, s_i \in \mathcal{S}\}$, $|\mathcal{C}| = J$ where r_i is the distance between node s_i and the collector, and d_{max} is the maximum transmission

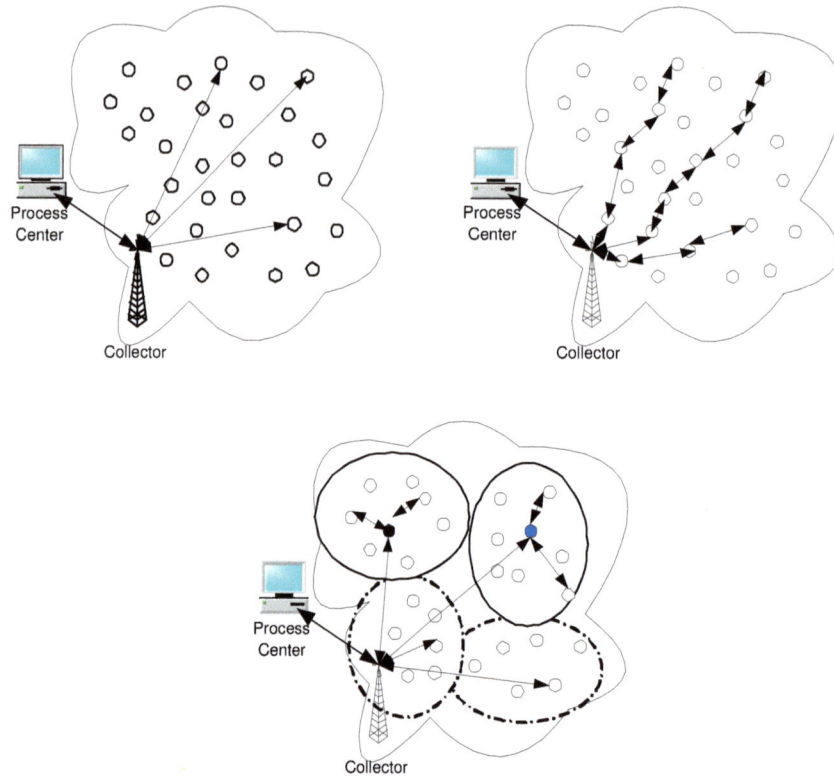

FIGURE 1. The three types of architecture of sensor networks.

range of a node, as the set of crucial sensors and the collector can only receive the packets from/through one node in this set. Therefore, the lifetime of the whole network is determined by the lifetime of set $\mathcal{C}$ which depends on the traffic generated and relayed by these nodes. In order to extend the network lifetime, we try to minimize the average energy consumption at these crucial nodes.

To achieve the goal of energy efficiency, it is necessary to identify the main sources that waste energy. The major sources of energy waste in sensor networks are: collision, overhearing, control packet overhead and idle listening. Due to the high traffic load at these crucial nodes, collision is the primary energy waste source and increases the latency as well.

3. System Model

Let us consider a wireless sensor network with K active transmitting crucial nodes, each using a slotted random access scheme [10]. Specially, node k transmits a length-N data packet, $\mathbf{d}_k(n) = \begin{bmatrix} d_{k,1}(n), & \cdots, & d_{k,N}(n) \end{bmatrix}^T$, during the n-th time slot. Assume that the collector (receiving node) uses an antenna array of M sensors. Within the n-th time slot, the data vector received by the m-th antenna can be modeled as,

$$(1) \qquad \mathbf{y}_m(n) = \sum_{k=1}^{K} a_{k,m}(n)\mathbf{d}_k(n) + \mathbf{v}_m(n),$$

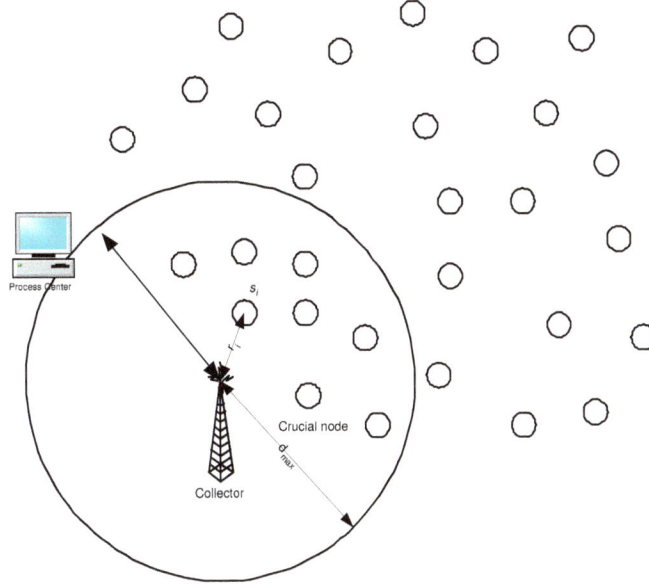

FIGURE 2. Crucial nodes in a multi-hop sensor networks.

where $a_{k,m}(n)$ denotes the channel gain (a real-valued Gaussian random variable) between the k-th transmitter (node) and the m-th antenna, vector $\mathbf{v}_m(n)$ is the real-valued addictive white Gaussian noise at the m-th antenna.

The total received data from all M antennas within the n-th time slot, collected in a $N \times M$ matrix, can be modelled as

$$
\begin{aligned}
\mathbf{Y}(n) &= \begin{bmatrix} \mathbf{y}_1(n), & \cdots, & \mathbf{y}_M(n) \end{bmatrix} \\
&= \sum_{k=1}^{K} \mathbf{d}_k(n)\, \mathbf{a}_k^T(n) + \mathbf{V}(n) \\
&= \mathbf{D}(n)\, \mathbf{A}(n) + \mathbf{V}(n),
\end{aligned}
$$

(2)

where $\mathbf{A}(n)$ is a $K \times M$ mixing matrix with $\mathbf{a}_k^T(n) = \begin{bmatrix} a_{k,1}(n), & \cdots, & a_{k,M}(n) \end{bmatrix}, k = 1, 2, \ldots, K$ as its rows; the $N \times K$ matrix $\mathbf{D}(n) = \begin{bmatrix} \mathbf{d}_1(n) & \cdots & \mathbf{d}_K(n) \end{bmatrix}$ contains K collided packages; $\mathbf{V}(n) = \begin{bmatrix} \mathbf{v}_1(n) & \cdots & \mathbf{v}_M(n) \end{bmatrix}$ is the noise matrix.

In a simple network using random access protocols, when a collision is detected, the received packets $\mathbf{Y}(n)$ are discarded and the system initiates a retransmission schedule. However, from (2) we know that the data matrix $\mathbf{Y}(n)$ contains information of all collided packets, hence, it should be exploited with the help of additional data transmission for collision resolution. We notice the fact that the reception diversity from the M-element receiving antenna array provides M copies of the transmitted packages $\mathbf{D}(n)$, each being scaled by a different channel gain vector (the m-th column of matrix $\mathbf{A}(n)$). Hence, we can resolve K transmission nodes' packages as long as $M \geq K$. Based on this observation, it is possible to propose a *collaborative approach* to combining the spatial diversity with the network assisted diversity to separate and extract information from received data packages in collision.

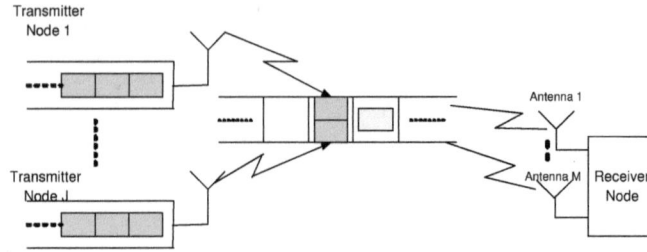

FIGURE 3. A random access, slotted wireless system with receiving antennas.

4. Collision Resolution through Signal Separation

4.1. Network Assisted Diversity Multiple Access. In principle, collision resolution is equivalent to signal separation. The network assisted diversity multiple access (NDMA) approach can be used for this purpose [6].

Let us consider the case when K transmission nodes collide in a given time slot n. The receiver has only one receive antenna. Therefore the received data vector $\mathbf{y}(n)$ consists of a mixture of K packages from different sources. That is,

$$\mathbf{y}_{NA}(n) = \sum_{k=1}^{K} a_k(n)\, \mathbf{d}_k(n) + \mathbf{v}(n) \; = \mathbf{D}(n)\, \mathbf{a}(n) + \mathbf{v}(n),$$

where the information $\mathbf{D}(n) = \begin{bmatrix} \mathbf{d}_1(n) & \cdots & \mathbf{d}_K(n) \end{bmatrix}$, and $\mathbf{a}(n) = \begin{bmatrix} a_1(n), & \cdots, & a_K(n) \end{bmatrix}^T$. Apparently, at least an additional $(K-1)$ snapshots of $N \times 1$ data vector are needed to resolve the $N \times K$ information matrix $\mathbf{D}(n)$. From a signal processing perspective, this problem may be solved if there is a method to create a Γ branch diversity with $\Gamma \geq K$ and collect Γ independent mixtures of the signals $\mathbf{d}_k(n)$. With the network layer knowledge, all transmission nodes are aware of the fact that a collision with multiplicity K occurred during the time slot n. Therefore, each of the K nodes will retransmit its packets $K-1$ more times in the next $K-1$ time slots. No other node will initiate a new transmission during these $K-1$ slots. With this collision detection and retransmission protocol, the receiver will receive a total K copies of the collided packets,

$$(3) \qquad \mathbf{Y}_{NA}(n) = \mathbf{D}(n)\,\mathbf{A}(n) + \mathbf{V}(n),$$

where $\mathbf{Y}_{NA}(n) = \begin{bmatrix} \mathbf{y}_{NA}(n) & \cdots & \mathbf{y}_{NA}(n+K-1) \end{bmatrix}$ is the $N \times K$ data mixing matrix, the $K \times K$ channel gain matrix is $\mathbf{A}(n) = \begin{bmatrix} \mathbf{a}(n) & \cdots & \mathbf{a}(n+K-1) \end{bmatrix}$, and the $N \times K$ noise matrix $\mathbf{V}(n) = \begin{bmatrix} \mathbf{v}(n) & \cdots & \mathbf{v}(n+K-1) \end{bmatrix}$.

If the mixing matrix $\mathbf{A}(n)$ of full rank is known or can be estimated, a simple linear inverse filtering solution can be used for data package separation,

$$(4) \qquad \hat{\mathbf{D}}(n) = \mathbf{Y}_{NA}(n)\,\mathbf{A}^{-1}(n).$$

Since only K time slots are required to construct a full rank mixing matrix $\mathbf{A}(n)$ for separating K collided packets, no slot is wasted and no throughput penalties incurred by the technique.

4.2. Space and Network Assisted Diversity Multiple Combining Access. Consider the similar case as in Section 4.1, but the receiver now is equipped with an array with M receiving antennas. Using the similar protocol, each of the K nodes will retransmit its information packet $L-1$ more times in the next $L-1$ slots (i.e., slots $n+1, \cdots, n+L-1$).

An example of this protocol for a collision of 3 nodes and 2 receive antennas is shown in the Figure 4.

$$(5) \qquad L = \lceil \frac{K}{M} \rceil$$

No other nodes will transmit a new packet in the next $L - 1$ slots. Assume that no other nodes will transmit a new packet in the next $L - 1$ slots. The proposed approach can resolve the packet collision during L time slots. Thus it is M times faster than NDMA to resolve colliding packets. The assumption holds when other nodes are notified by using a busy tone signal as stated in subsection 4.3. Only the colliding nodes will retransmit their packets in the next $L - 1$ slots according to the in-band busy signal while the other nodes will remain deferred within the $L - 1$ slots. Considering that both NDMA and proposed approach assume that no other active users transmit packets during the resolution period, the new approach is able to provide more opportunities for other deferred nodes to access channels in the next time slots. Since they are less likely to enter the back off waiting stage, the proposed approach achieves higher throughput and fairness than NDMA. The receiver will receive a total $L \times M$ copies of the collided packets with these conventions.

$$(6) \qquad \mathcal{Y}(n) = \mathbf{D}(n)\,\mathcal{A}(n) + \mathcal{V}(n),$$

where data $\mathcal{Y}(n)$ is $N \times (ML)$ matrix, the channel gain $\mathcal{A}(n)$ is $K \times (ML)$ matrix and noise $\mathcal{V}(n)$ is $N \times (ML)$ matrix,

$$\mathcal{Y}(n) = \begin{bmatrix} \mathbf{Y}(n) & \cdots & \mathbf{Y}(n+L-1) \end{bmatrix}$$
$$\mathcal{A}(n) = \begin{bmatrix} \mathbf{A}(n) & \cdots & \mathbf{A}(n+L-1) \end{bmatrix}$$
$$\mathcal{V}(n) = \begin{bmatrix} \mathbf{V}(n) & \cdots & \mathbf{V}(n+L-1) \end{bmatrix}$$

Equation (6) represents a classical source separation problem. If the mixing matrix $\mathcal{A}(n)$ is known or can be estimated, the maximum likelihood estimation of the transmitted packets is

$$(7) \qquad \hat{\mathbf{D}}(n) = \arg\min_{\mathbf{D}} \| \mathcal{Y}(n) - \mathbf{D}\,\mathcal{A}(n) \|_F^2$$

where $\|\cdot\|_F$ represents the Frobenius norm, and $\mathbf{D}$ takes all possible finite values. Since $ML \geq K$, the $rank\,(\mathcal{A}(n)) \geq K$. The ZF and MMSE solution for the desired data [12] can be gotten.

$$(8) \qquad \hat{\mathbf{D}}_{ZF} = \mathcal{Y}(n)\mathcal{A}^H(n)(\mathcal{A}(n)\mathcal{A}^H(n))^{-1}$$

$$(9) \qquad \hat{\mathbf{D}}_{MMSE} = \mathcal{Y}(n)\mathcal{A}^H(n)(\mathcal{A}(n)\mathcal{A}^H(n) + \sigma^2\mathbf{I})^{-1}$$

4.3. The Signal Separation. Generally, a guard period is needed at the beginning of each slot in a slotted network [13]. Every node listens to the carrier during this guard period, and transmits its data after this period. One can use such a guard period to inform nodes of the collision with a busy signal. The receiver has to discriminate all the active transmitting nodes. There are 2^J different possibilities in a J-transmitting-node system. A unique ID sequence for each node contained in the packet is required in order to enable the receiver to uniquely identify all the active transmitting nodes. Assume that the first

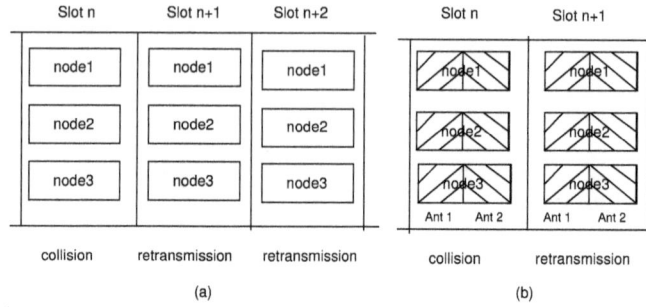

FIGURE 4. Packet collision and retransmission with 3 transmission nodes and 1 to 2 receiving antennas. With spatial diversity at reception, only 2 transmissions are required to resolve collision among 3 transmission nodes in (b).

Q symbols of each packet of node k is the ID sequence, that is $\bar{\mathbf{d}}_k = [\mathbf{d}_k(n)]_{1:Q}$. The corresponding received data is,

$$(10) \qquad \bar{\mathbf{y}}_m(n) = \sum_{k=1}^{K} a_{k,m}(n)\,\bar{\mathbf{d}}_k + \bar{\mathbf{v}}_m(n)$$

where $\bar{\mathbf{y}}_m(n) \triangleq [\mathbf{y}_m(n)]_{1:Q}$, and $\bar{\mathbf{v}}_m(n) \triangleq [\mathbf{v}_m(n)]_{1:Q}$. From (10), the estimation of $a_{k,m}(n)$ from the received data $\bar{\mathbf{y}}(n)$ is related to a least square (LS) problem [11]. Assume that the ID sequences are orthogonal to each other.

$$\bar{\mathbf{d}}_k^T \bar{\mathbf{d}}_l = \{ \begin{array}{cc} 0 & k \neq l \\ 1 & k = l \end{array} .$$

the matched-filter output $z_{k,m}(n)$ associated with $\bar{\mathbf{d}}_k$ is

$$(11) \qquad z_{k,m}(n) = \bar{\mathbf{d}}_k^T \bar{\mathbf{y}}_m(n) = a_{k,m}(n) + \bar{\mathbf{d}}_k^T \bar{\mathbf{v}}(n),$$

Therefore vector $\mathbf{z}_m(n) = [z_{1,m}(n), \cdots, z_{K,m}(n)]^T$ forms the sufficient statistic for estimating the node gain $\mathbf{a}_m(n) = [a_{1,m}(n), \cdots, a_{K,m}(n)]^T$ for the mth antenna.

$$(12) \qquad \hat{\mathbf{a}}_m(n) = \mathbf{z}_m(n)$$

5. Goodput and Delay Analysis

5.1. Goodput Analysis. It is instructive for us to view the traffic in the channel as a flow of collision resolution periods or epochs. An epoch includes one or several consecutive channel slots that are dedicated for the transmission (including the initial transmission and retransmissions) of the data packets from the nodes who are active at the beginning of the epoch. The idle slots, during which no data are transmitted, also compose epochs called idle epochs, which only include one slot. Correspondingly, we call those epochs, during which some packets are under transmission, busy epochs. The length of a busy epoch is the number of time slots the channel takes to serve the currently active transmission nodes.

The epoch length is a random variable depending on the number of the active transmission nodes at the beginning of the epoch. If denote by P_{emp} the probability of a transmission node's buffer being empty at the beginning of an epoch, then binomial expressions for probability of the length l epoch busy or idle can be obtained,

$$(13) \qquad P_{busy}(k) = \binom{J}{k}(1 - P_{emp})^k P_{emp}^{J-k}$$

$$(14) \qquad P_{idle}(k) = \begin{cases} P_{emp}^J & k = 1 \\ 0 & \text{otherwise} \end{cases}$$

where $k = 1, 2, \cdots, J$ is the number of active transmission nodes and J is the total number of transmission nodes in the network.

Let us define the goodput as

$$R = \frac{\text{average length of busy epoch} \cdot (1 - P_e)}{\text{average length of (busy or idle) epoch}}$$

We can obtain

$$(15) \qquad R = \frac{\sum\limits_{k=1}^{J} k \binom{J}{k}(1 - P_{emp})^k P_{emp}^{J-k} \cdot (1 - P_e(k))}{\sum\limits_{k=1}^{J} k \binom{J}{k}(1 - P_{emp})^k P_{emp}^{J-k} + 1 \cdot P_{emp}^J}$$

where $P_e(k)$ is the bit error rate for active transmission node k.

5.2. Delay Analysis. From the viewpoint of a particular transmission node, two types of epochs (see Figure 5) can be distinguished: relevant epochs, in which a data packet belonging to this node is being transmitted, and irrelevant epochs, in which no packet belonging to this node is being transmitted. The lengths of two types of epochs, denoted by l_r and l_i, obey different distributions,

$$(16) \qquad P_{l_r}(L) = \binom{K-1}{J-1}(1 - P_{emp})^{K-1} P_{emp}^{J-K}, \qquad 1 \le K \le J$$

$$(17) \qquad P_{l_i}(L) = \begin{cases} P_{emp}^{J-1} + (J-1)(1 - P_{emp})P_{emp}^{J-2}, & K = 1 \\ \binom{K}{J-1}(1 - P_{emp})^K P_{emp}^{J-K-1}, & 1 < K \le J-1 \end{cases}$$

where $L = K$ in NDMA scheme, and for space and network combining approach L is chosen according to (5).

Denote q_m as the number of data packets in the buffer of a user at the beginning of the mth epoch. The sequence constitutes a Markov chain.

$$(18) \qquad q_{m+1} = \begin{cases} q_m - 1 + v(q_m) & q_m > 0 \\ v(q_m) & q_m = 0 \end{cases}$$

where $v(q_m)$ be the number of data packets arriving during the mth epoch.

The probability generating function of q_m is

$$(19) \qquad Q_m(z) = \sum_{k=0}^{\infty} Pr q_m = k z^k = E[z^{q_m}]$$

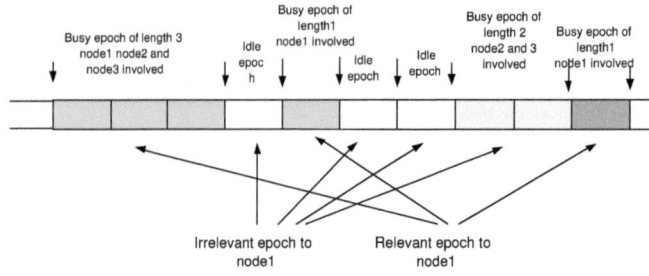

FIGURE 5. Epoch flow and types of epochs (busy epoch and idle epoch; relevant epoch and irrelevant epoch

and the steady state of $Q_m(z)$ is $Q(z) = \lim_{m \to \infty} Q_m(z)$.

$$F(z) = \lim_{m \to \infty} E[z^{v(q_m)}|q_m = 0]$$

$$G(Z) = \lim_{m \to \infty} E[z^{v(q_m)}|q_m > 0]$$

If the buffer is fed by a Poisson source with rate λ, the steady-state probability generating function $Q(z)$ is

$$Q(z) = P_{emp} \frac{zF(z) - G(z)}{z - G(z)}$$

and

$$F(z) = \sum_{k=1}^{J} e^{k(z\lambda - \lambda)} P_{lr}$$

$$G(Z) = \sum_{k=1}^{J-1} e^{k(z\lambda - \lambda)} P_{li}$$

Let $z = 1$ in equation 19, we can find the relationship between P_{emp} and λ, P_{emp} is the unique solution in $[0, 1]$ of the equation

$$(20) \qquad \lambda P_{emp}^{J} + (1 + \lambda J)P_{emp} - (1 - \lambda J) = 0$$

A transmission node's buffer can also be modeled as an M/G/1 queue with vacation, in which the relevant epoch and irrelevant epoch play the role of the service time and vacation time respectively. According to the property of the M/G/1 queue with vacation, the average system delay (including the waiting time in the buffer and the transmission time in the channel) for a data packet can be expressed as

$$(21) \qquad D = \bar{l}_r + \frac{\lambda \bar{l}_r^2}{2(1 - \lambda \bar{l}_r)} + \frac{\bar{l}_i^2}{2\bar{l}_i}$$

where packet arrival is a Poisson process with packet arrival rate λ, $\bar{l}_r, \bar{l}_r^2, \bar{l}_i$ and $\bar{l}_i^2$ are the first and second moments of the relevant epoch and irrelevant epoch, respectively, and can be computed from their distribution.

6. Energy Consumption and Lifetime of the Network

For a crucial sensor node $s_i \in \mathcal{C}$, use $E_{i,init}$ to denote the initial energy of the node, $E_{i,s}$ to denote the energy needed to sense data (one bit), $E_{i,p}$ to denote the energy needed

TABLE 1. Assumed Parameter for Crucial Nodes at Sensor Networks

Parameter	Value
Fade margin and Pass loss $P_l\alpha$	70dB
Thermal noise at collector N_t	-174dBm
Receiver noise at collector N_r	10dB
Signal bandwidth W	1MHz
Data rate R	1Mbps

for data processing (one bit), and $E_{i,Rx}/E_{i,Tx}$ to denote the energy needed for receiving and transmitting the one bit data, respectively.

Since different collision separation methods are utilized in the collector, assume the $E_{i,s}$, $E_{i,p}$ are the same for all methods. The energy needed to receive a bit $E_{i,Rx}$ accounts for the receiver electronics energy dissipation. The energy needed to transmit a bit $E_{i,Tx}$ consists of two parts: the energy dissipation of the transmitter electronics $E_{i,txe}$, and the RF transmit energy $E_{i,RF}$.

The energy per bit to noise ratio at the receiver is [14]

$$\frac{E_b}{N_0} = \frac{P_{RF}}{P_l\alpha} \cdot \frac{1}{WN_tN_r}$$

where P_{RF} is the RF transmission power, P_l is the large scale path loss, α is the average attenuation factor due to fading, W is the signal bandwidth, N_t is the thermal noise and N_r is the noise at the receiver known as the noise figure. In general, $P_l \propto \frac{1}{4\pi d^k}, 2 \le k \le 4$.

The transmit power $P_{i,RF}$ can be written as

$$(22) \qquad P_{i,RF} = P_l\alpha WN_tN_r\frac{E_b}{N_0}$$

The assumed parameters are given in Table 1.

In the sensor network, collision happens when the number of the active crucial nodes is greater than the number of the antennas at the collector ($K > M$ at antenna array scheme and $K > 1$ at single antenna scheme), collided packets will be retransmitted for collision separation, and the retransmission time depends on K. The average RF energy required to transmit a bit for each node is a function of the number of the active nodes, one defines it as $E_{RF}(K)$

$$(23) \qquad E_{RF}(K) = \frac{1}{K}KP_{RF} \cdot T_k = P_{RF} \cdot T_k$$

where $T_k = KT$ in NDMA scheme, and for SNDC approach $T_k = LT$. Then

$$E_{Tx}(K) = E_{RF}(K) + E_{txe}$$

Define the probability of a node's buffer being empty at the beginning of an epoch as P_{emp} [10], the probability of that K nodes are active is $P_a(K)$

$$P_a(K) = \binom{J}{K}(1 - P_{emp})^K P_{emp}^{J-K}$$

The average energy consumption for transmitting per bit per node is

$$E_{Tx} = \sum_{K=0}^{J} E_{Tx}(K)P_a(K)$$

Assume the average traffic rate in the sensor network is λ, λ_g and λ_{re} are the rate of traffic generated and relayed by s_i, the traffic generated by the crucial node is assumed to be the same for all nodes in the network and $\lambda_g = \lambda$, the traffic relayed by the crucial node is

$$\lambda_{re} = \frac{N_{total} - J}{J}\lambda$$

The power of node s_i is

$$
\begin{aligned}
P_i &= (E_{i,s} + E_{i,p} + E_{i,Tx})N\lambda_{g,i} \\
&+ (E_{i,Rx} + E_{i,p} + E_{i,Tx})N\lambda_{re,i} \\
&\geq E_{i,Tx}N(\lambda_{g,i} + \lambda_{re,i}) + E_{i,Rx}N\lambda_{re,i} \\
&= P_{i,Tx/Rx}
\end{aligned}
$$

(24)

The bound of the lifetime for node s_i is

$$t_i \leq \frac{E_{i,init}}{P_{i,Tx/Rx}}$$

The lifetime of the network is

(25)
$$t_{net} = \max\{t_i|s_i \in \mathcal{C}\}$$

7. Simulation Results

To observe the performance of the space and network assisted diversity multiple access method, it is compared with the pure NDMA method through simulation. Consider a slotted data communication system, the total number of node in the network is $N_{total} = 200$, the number of crucial nodes is $J = 32$, and the nodes' ID sequences gold code with code length $Q = 31$. The number of the receive antenna is $M = 2$. The transmission packets are fixed length of $N = 424$ bits (equal to the length of an ATM cell). Assume The initial energy at each crucial node $E_{init} - 6J$ [**15**].

The simulation are carried under four SNR cases:

- 5 dB;
- 10 dB;
- 20 dB;
- 30 dB;

Under each scenario, the number of the active transmission nodes in the system changes.

The simulation results of goodput versus total traffic load according to (15) are shown in Figure 6. Figure 6 shows the goodput versus total traffic load λJ corresponding to SNR from 5dB to 30dB. We present two results: the SNDC and the NDMA. The performance the of SNDC is better than NDMA only, especially in low SNR. From the comparison between the analytical and simulation results, the simulation results are in good agreement with the analytical expressions.

The delay performance is shown in Figure 7 as a function of the traffic load. The analytical and simulation results demonstrate that the delay performance of the NDMA and space and network combining is much better than TDMA and pure ALOHA. We may see the lowest latency property of the space and network combining approach comparing with other approaches.

In Figure 8 and 10, one can see the energy consumption (transmission and receiving) increase with the traffic load, and energy consumption of the SNDC with MMSE separation method is much smaller than the others, especially when the lower BER is want to be achieved.

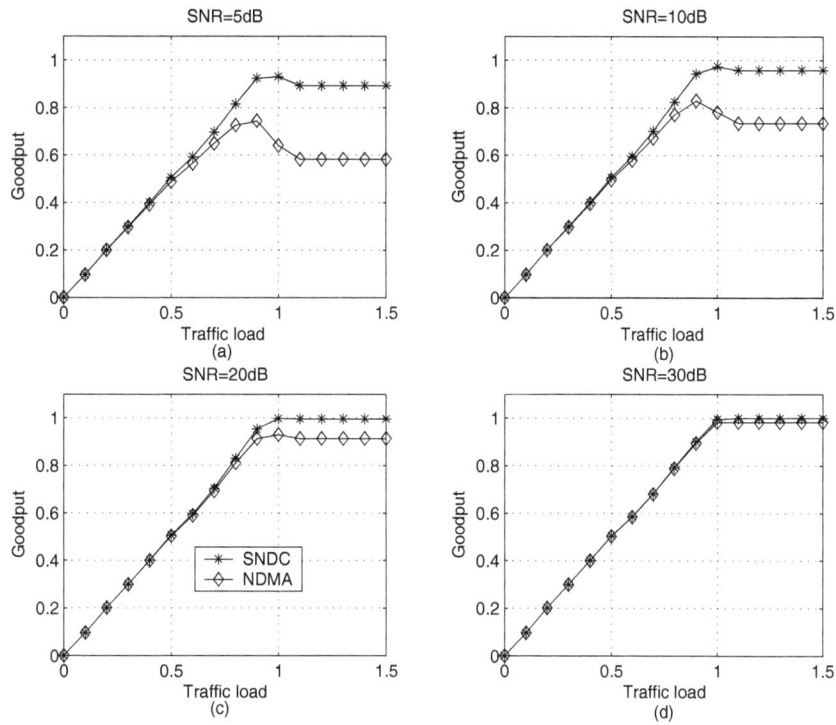

FIGURE 6. Simulation result: goodput vs. traffic load with between different collision resolution approaches. System parameters: $J = 32$, $M = 2$ (a) 5dB (b) 10dB (c) 20 dB (d) 30 dB

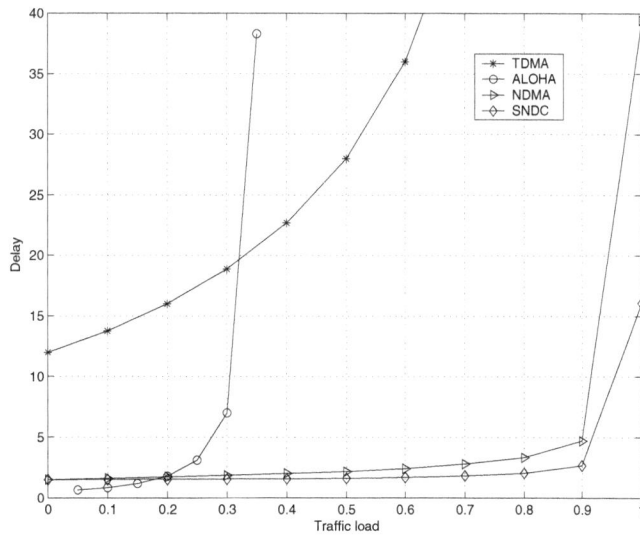

FIGURE 7. Performance comparison (delay vs. traffic load) among different approaches. (analytical results)

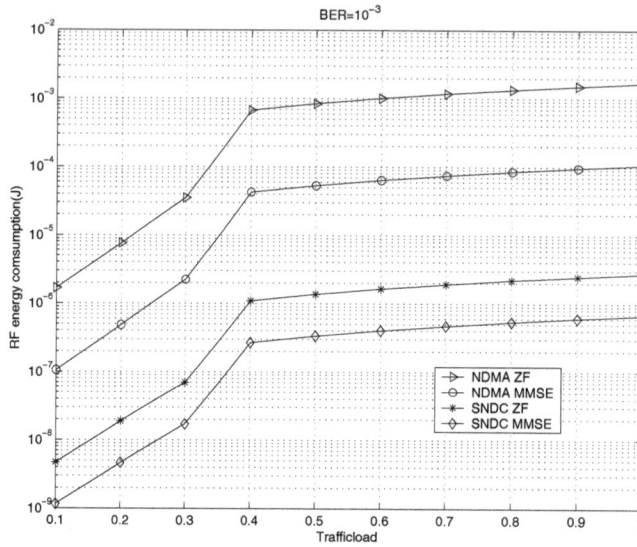

FIGURE 8. The energy consumption vs. traffic load using various of transmission methods (BER=10^{-3})

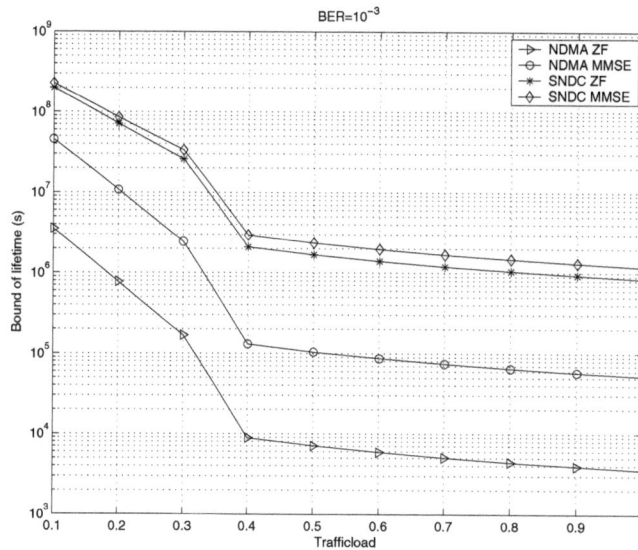

FIGURE 9. The lifetime vs. traffic load using various of transmission methods (BER=10^{-3})

Figure 9 and 11 show the simulation results of the network lifetime vs. traffic load. The lifetime decrease when traffic load increase, and the lifetime of the SNDC with MMSE separation method is much larger than the others, especially when the lower BER is want to be achieved.

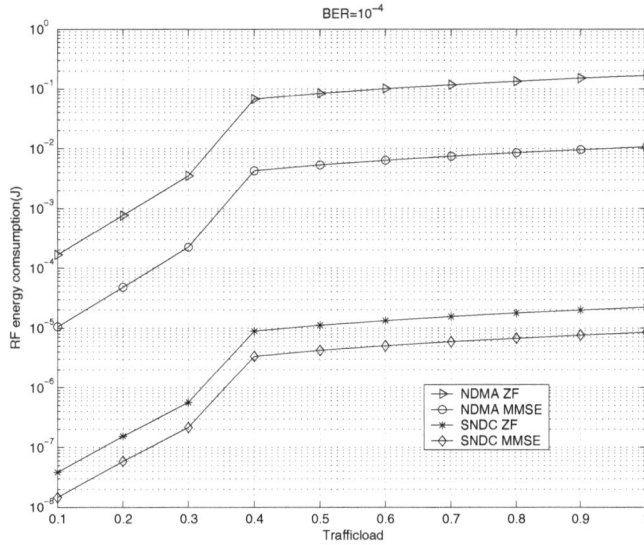

FIGURE 10. The energy consumption vs. traffic load using various of transmission methods (BER=10^{-4})

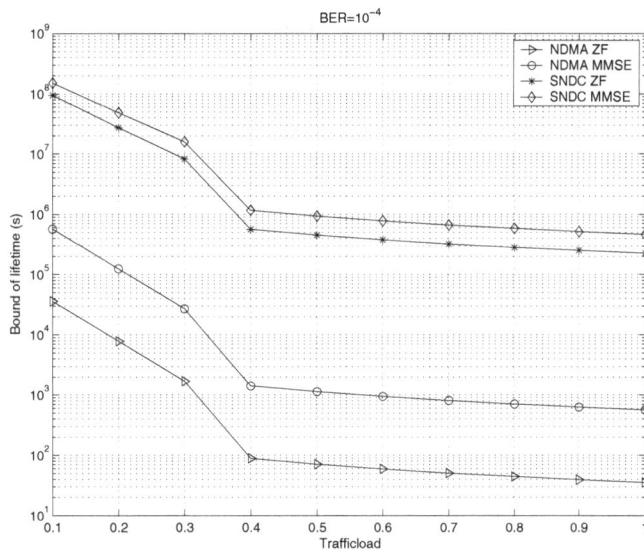

FIGURE 11. The lifetime vs. traffic load using various of transmission methods (BER=10^{-4})

8. Conclusions

A new cross-layer design approach for network diversity is presented in this chapter. The proposed collision resolution scheme to multiple access in wireless sensor network is based on the combination of space and network assisted diversity. The diversity is fully exploited by the receiver for collision resolving and package separation. It is demonstrated that this approach can extract the useful information from collided packets. Compared the the network-only assisted diversity approach with proposed space and network diversity, the latter improves the network performance in terms of goodput and delay. The space and network combining approach also provides smaller average energy consumption and longer network lifetime.

Bibliography

[1] D. P. Bersekas and R. Gallager, *Data Networks*, Upper Saddle River, NJ: Prentice Hall, (1992).

[2] A. S. Tanenbaum., *Computer Networks*, 3rd ed. Upper Saddle River, NJ: Prentice Hall, (1996).

[3] A. J. Paulraz and B. Ng, "Space-time modems for wireless personal communications," IEEE Personal Communications, vol. 5, pp. 56-84, (1998).

[4] A. J. Paulraz and C. B. Papadias, "Space-time processing for wireless communications," IEEE Signal Processing Magazine, pp. 49-93, (1997).

[5] S. Verdu, *Multiuser Detection*, Upper Saddle River, NJ: Cambridge Univ. Press, (1998).

[6] M. K. Tsatsanis, R. Zhang, and S. B. Banerjee, "Network-assisted diversity for random access wireless networks," IEEE Transactions on Signal Processing, vol. 48, no. 3, pp. 702-711, (2000).

[7] W. R. Heinzelman, A. Chandrakasan, and H. Balakrishnan, "Energy-efficient communication protocol for wireless microsensor networks," in Proc. International Conference on System Sciences, Hawaii, pp. 1-10, (2000).

[8] M. Bhardwaj and A. P. Chandrakasan, "Bounding the lifetime of sensor networks via optimal role assignments," in Proc. IEEE INFOCOM: The Conference on Computer Communications, vol. 3, pp. 1587-1596, (2002).

[9] L. Liu and H. Ge, "Space and network assisted diversity for linear mmse collision separation in wireless sensor networks," in Proc. Conference on Information Sciences and Systems, vol. 1, Baltimore, MD, pp. 961-966, (2005)

[10] L. Liu and H. Ge, "Space and network assisted diversity for cross-layer design in wireless networks," in Proc. Conference on Information Sciences and Systems, vol. 2, Princeton, NJ, 2004, pp. 961-966, (2004)

[11] L. Liu, S. Gujjula, and S. M. Kuo, "Multi-channel real time active noise control system for infant incubators," in Proc. IEEE Engineering in Medicine and Biology Society, pp. 935-938, (2009).

[12] S. Haykin, *Adaptive Filter Theory*, Upper Saddle River, NJ: Prentice Hall, (1996).

[13] A. Ebner, H. Rohling, M. Lott, and R. Halfmann, "Decentralized slot synchronization in highly dynamic ad hoc networks,' in Proc. International Symposium on Wireless Personal Multimedia Communications, Honolulu, Hawaii, pp. 27-30, (2002).

[14] E. Shih, S. Cho, and N. Ickes, "Physical layer driven protocol and algorithm design for energy-efficient wireless sensor networks," in Proc. ACM International Conference on Mobile Computing and Networking, Rome, Italy, pp. 272-286, (2001)

[15] J. Zhu and S. Papavassiliou, "On the energy-efficient organization and lifetime of multihop sensor networks," IEEE Communications Letters, vol. 7, pp. 537-539, (2003).

Send Orders of Reprints at reprints@benthamscience.net

CHAPTER 7

Intelligent Space:
A Platform for Integration of Robot Technology

Takeshi Sasaki and Hideki Hashimoto

Institute of Industrial Science

The University of Tokyo

Tokyo, Japan

ABSTRACT. Latest advances in network sensor technology and state of the art of mobile robotics and artificial intelligence research can be employed to develop autonomous and distributed monitoring systems. We have been developing "Intelligent Space (iSpace)," which is a space with ubiquitous sensors and actuators. Most of intelligent system interacts with human in a passive space, but iSpace, a space that contains human and artificial systems, is an intelligent system itself. Specific tasks, which cannot be achieved by the iSpace, are accomplished by using the artificial systems. For examples, iSpace utilizes computer monitors to provide information to the human, and robots are utilized to provide physical services to the human as physical agents. In this chapter, we summarize the present state of iSpace and describe the future work from the viewpoint of system integration. We introduce our ongoing researches on essential functions of iSpace – "observation," "recognition" and "actuation."

1. Introduction

Intelligent Space (iSpace) is an environmental system, which has multiple distributed and networked sensors and actuators. iSpace observes the space using the distributed sensors, extract useful information from the obtained data and provide various services to users. This means their essential functions are "observation," "recognition" and "actuation." This type of spaces is also referred to as smart space, intelligent environment, etc. and recently there is a growing number of research work [8].

The ultimate goal of Intelligent Space project is to accomplish an environment that comprehends human's intentions and satisfies them. It appears that such a system is hardly achieved, since a huge number of functions should be prepared and human-like intelligence is required. Even though such a complete system cannot be achieved immediately, we are certain that a useful system can be achieved with current technology by proper system integration.

Fig. 1 shows the concept of iSpace, which is able to support human in informative and physical ways. In iSpace, not only sensor devices but also sensor processing intelligence is distributed in the space because it is necessary to reduce the network load in the large-scale network and it can be realized by processing the raw data in each sensor node before collecting information. We call the sensor nodes distributed in the space DINDs (Distributed Intelligent Network Device). A DIND consists of three basic components: sensors, processors and communication devices. The processors deal with the sensed data and extract useful information about objects (type of object, position, etc.), users (identification, posture, activity, etc.) and the environment (geometrical shape, temperature, emergency, etc.). The network of DINDs can realize the observation and understanding of the events in the whole space. Based on the extracted and fused information, actuators such as displays or projectors embedded in the space provide informative services to users.

In iSpace, mobile robots are also used as actuators to provide physical services to the users and for them we use the name mobile agents. A mobile agent can utilize the intelligence of iSpace. By using distributed sensors and computers, the mobile agent can operate without restrictions due to the capability of on-board sensors and computers. Moreover, it can understand the request from people and offer appropriate services to them.

In this chapter, Section 2 presents the current configuration of iSpace and technologies for efficient development and integration of iSpace system. The implemented system is also introduced. Section 3, 4, and 5 present ongoing researches on essential functions of

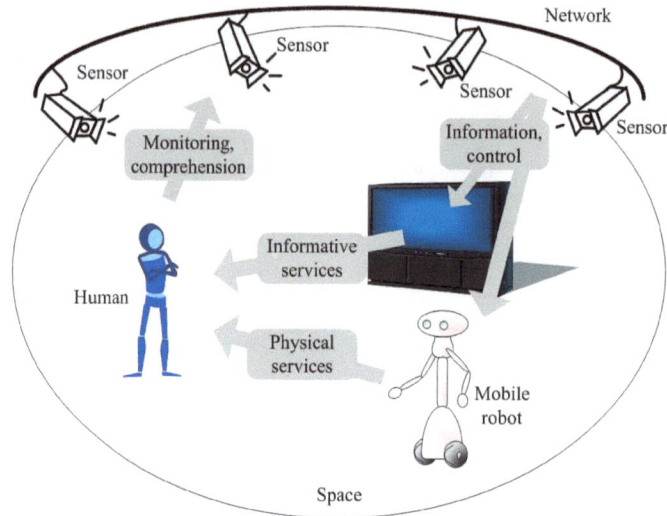

FIGURE 1. Concept of Intelligent Space.

iSpace – "observation function," "recognition function" and "actuation function," respectively. Finally, conclusion and future work are given in Section 6.

2. Design and Implementation of Intelligent Space

Configuration. This section focuses on the structure of iSpace, at which information from all DINDs are fused and the combined information is utilized by the applications to perform informative and physical actuation. Fig. 2 shows the configuration of iSpace. It consists of four layers - the sensor node layer, the basic information server layer, the application layer and the actuator layer.

The information obtained by each sensor node (DIND) is first sent to the corresponding basic information servers along with the reliability of the measurement that is used for information fusion. For example, geometrical shape of the environment and positions of targets are sent to the map server and the position server respectively. That is, each sensor node sends the information that can be obtained. The basic information server combines the information based on the reliability and, if necessary, sends the result back to the sensor nodes. The returned information can be used for the next measurement in the sensor node. The servers can share the information combined in each server using the database. The applications request the information they need from basic information servers. Based on the received information, the applications send the commands to the actuators and realize various services to the users. For example, an information display system would get the map of the space and the position of the humans from the map and position server respectively, find an empty area in front of the human and project an image using a projector. The applications only need to know the basic information server's address to get information, so they do not need to connect directly to the sensor that observes the target object. In addition, with this structure it is possible to add, remove or replace the sensor devices or sensor nodes without any change in the application program.

System Development and Integration.

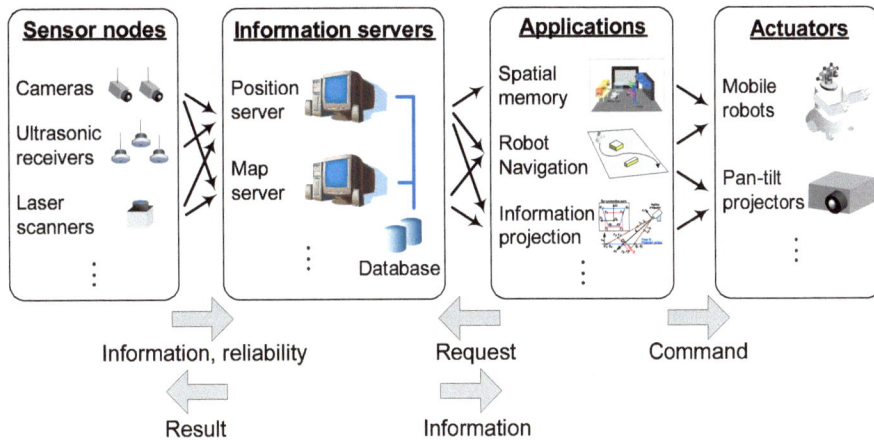

FIGURE 2. Configuration of Intelligent Space.

Algorithm evaluation support tools. With the fast development of sensor technology, lower cost and more types of sensors are available in iSpace now. Currently, new algorithms are tested through experiments. Experiments are the best way to evaluate the developed system. However, experiments require a lot of time and effort and are limited by the structure of the space. Therefore three algorithm evaluation support tools, Environment Designer, Sensor Arranger and Simulation Runner, are developed to replace experiments with computer simulations [35].

Environment Designer is used to create various environments. Several different environment elements including walls, tables and chairs are provided to build an indoor environment. The created environments can be used in Sensor Arranger which is used to arrange various sensors in the environment. We offer five kinds of sensors in our platform and the sensor parameters (e.g. observable range and angle of a laser range finder) can be set for each sensor. After creating environment and arranging sensors, Simulation Runner executes distributed sensor simulation through communication with external DIND processing part. Because the motivation of this program is not only to aid the process of research and experiments, but also to help new comers to create their specific iSpace, we also offer a GUI for manipulation of these tools.

Fig. 2 shows sensor arrangement evaluation and simulation of mobile robot navigation using the developed tools.

Component based integration using RT-Middleware. iSpace should have flexibility and scalability so that we can easily change the arrangement of embedded devices and switch applications depending on the size of the space, technological advances, etc.

A solution to this problem is a modularization. In module or component based systems, independent elements (modules or components) of functions of the systems are first developed and the systems are then built by combining the modules. The modularization increases maintainability and reusability of the elements. Moreover, flexible and scalable system can be realized since the system is reconfigured by adding or replacing only related components. In order to receive the benefit of modularization, it is necessary to ensure the connectivity between components. This means that the interface between components should be standardized. Considering the cooperation of components that are developed by

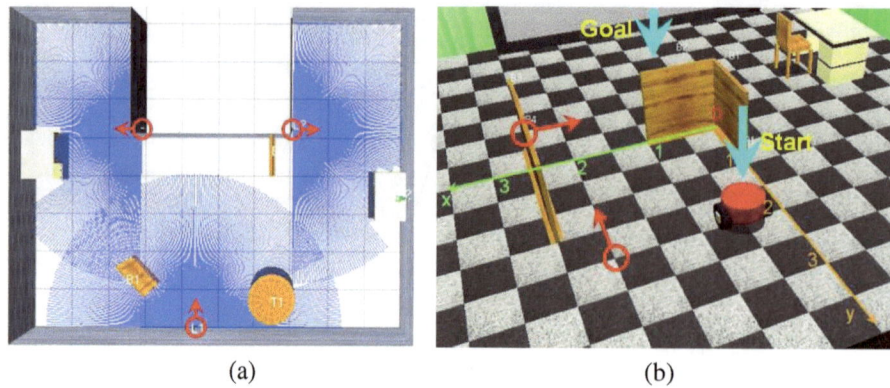

(a)   (b)

FIGURE 3. Snapshot of development support tools (a) evaluation of arrangement of laser range finders (b) simulation of obstacle avoidance.

various manufactures, international standardization is desirable. Recently, OMG (Object Management Group) adopted the RTC (Robotic Technology Component) Specification that defined a component model for development of robotic systems.

Although some platforms that support component based system integration have been developed [6, 7], we use OpenRTM-aist [2] since it is a middleware that complies with the standard. It also includes a template code generator which makes a source code of a component from the specification of the component (e.g. number of I/O ports, etc.) and a system design tool which provides graphical user interface to change the connection of components and start/stop the system. As mentioned in the previous section, since iSpace consists of four layers, we made RT-Components based on the functions of these layers [27].

Implementation. Fig. 2 shows the implemented iSpace in Hashimoto Lab. Our laboratory room, which is about 5m × 8m, is used for the iSpace. The present configuration involves eight pan-tilt-zoom CCD cameras handled by 4 sensing nodes (PCs), several laser range finders, an ultrasound positioning system, a 3D-camera, a wireless sensor network system and two mobile robots. Moreover, the iSpace has a large size screen, a pan-tilt projector and speakers for presenting information to the users of the space. All the modules are connected through the local area network. Also, for achieving appropriate conditions for the operation of cameras, the lighting in the space can be easily adjusted.

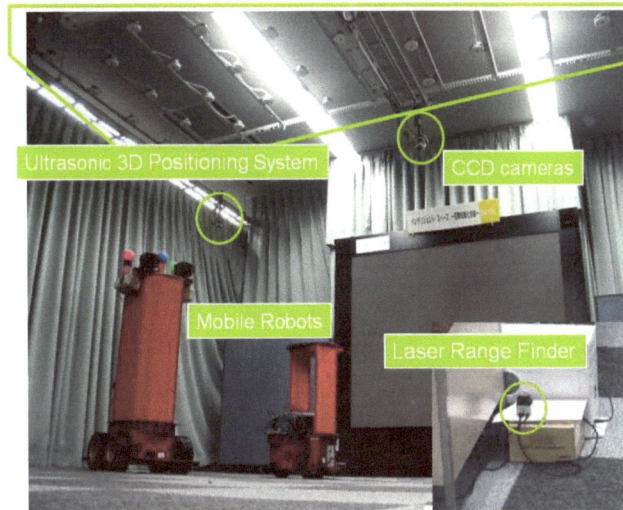

FIGURE 4. Experimental environment.

1) CCD Cameras

In the iSpace, CCD cameras are used as sensors for tracking the objects in the space. The cameras are placed so that the whole area of the room is covered. The placement of the cameras can also be optimized to expand the viewable area of the cameras, which was a subject of our previous research [1]. In our currently used system, the human and robot tracking is done by background subtraction and using color information [19]. The robot tracking makes use of color markers positioned on top of the robot for identification. Using the information from all cameras the position of both the humans and robots can be reconstructed. The tracking software also provides a GUI for easy operation.

2) Ultrasound positioning system

The ultrasound positioning system in the iSpace is used to obtain the 3D position of objects in the space. It consists of small size transmitters and 96 receivers positioned on the ceiling and connected to the control unit. The transmitters send an ultrasound signal, which is detected by the receivers from which the position of the transmitter can be calculated. In order to obtain the information form the positioning system the nodes on the network access a network server connected with the system.

3) Laser range finders

The current iSpace uses multiple laser range finders to measure the 2D position of mobile robots, humans, and other objects. Also, laser range finders are used as on-board sensors for mobile robots. The details of our laser range finder based tracking method and evaluation of the method are described in section 3.

4) Wireless Sensor Network

A wireless sensor network is also constructed in the iSpace. As one of its application, 3-axis acceleration sensors are connected to the nodes of the sensor network. These acceleration sensors are attached to objects in the iSpace so that it is possible to monitor the interaction among these objects and their users.

5) Mobile robots

In the iSpace we currently use two mobile robots. A PC is mounted on each mobile robot in order to provide processing of data from the sensors mounted on robots and for communication with the iSpace via wireless LAN. For the detection of the robot position information, cameras, ultrasound system, laser range finders and robot wheel encoders are used. Based on the obtained position, tracking and position control of the robot are performed. The robots also have on-board sensors including cameras and laser range finders. The observations from both distributed and on-board sensors are also used to detect the humans and obstacles in the space, which is in turn used for planning the path of the robots.

3. Observation Function

Object Tracking Using Distributed Laser Range Finders. Position information is one of the most basic information for iSpace. There exist various systems for indoor tracking of humans or other objects. The characteristics of such systems vary greatly with the type of sensors used [13]. The types of tracking systems can be roughly divided to systems where each sensor can track the objects independently and can be used either alone or in combination with other sensors (for example, cameras - especially stereo cameras, or laser range finders), and systems where multiple sensors are used together to give one tracking system. Although the latter tracking systems often offer more accurate measurements, they require complex installation. So we mainly use single sensors including CCD cameras and laser range finders for applications in iSpace and the ultrasound positioning system is utilized to validate the results in the tracking experiment.

Cameras are inexpensive and easy to install, which is the reason why they have frequently been employed for indoor tracking in research works on intelligent environments [14,18,25]. As mentioned above, we have also developed and used a camera based system for tracking both humans and robots in iSpace. However, both in our and other researches, it has proven hard to obtain a robust and easy to use system for tracking both humans and robots using only cameras. This is due to the complexity of the observed environments, effects of changing illumination, etc. Laser range finders have recently been frequently used for a variety of sensing tasks. This is probably due to the appearance of relatively low priced eye-safe laser range finder devices. They are also considered to be one of the most useful sensors in smart environment applications since it has high accuracy and can be used for both tracking and mapping. In this subsection, tracking processes in laser range finders are presented in more detail.

A scan from the laser range finder is taken at each time step. The obtained scan is then processed through several steps in order to obtain the position of a human or robot in the scan. The flow of the tracking process is shown in Fig. 3. It consists of background subtraction, clustering, data association and tracking, which are performed separately on each distributed sensor. The results are then sent to a fusion server, where the data from all sensors are combined. Each of these steps is described in the following.

1) Background subtraction

Background subtraction is processes of determining which parts of the scan are due to static objects (the background), and which come from reflections from moving objects, in our case humans or robots (the foreground).

The background can be easily determined by taking several scans while there are no moving objects in the space and taking their average. The foreground can then be extracted from the scan by comparing with the learned background: the parts of the scan that differ from the background more than a given threshold are marked as foreground.

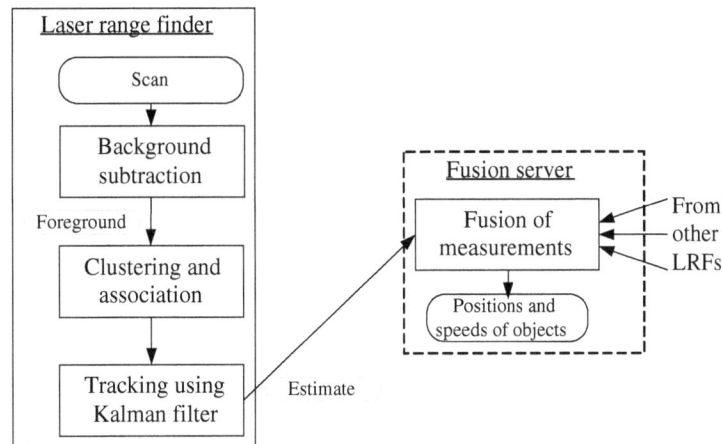

FIGURE 5. Flow of the tracking process.

2) Clustering and data association

The scan points in the foreground are clustered using a nearest neighbor classifier based on the Euclidean distance. This divides the foreground to a number of clusters, where each cluster belongs to only one of the tracked objects. Clusters with a small number of scan points are discarded, which effectively eliminates the noise that can be present in the measurement.

In order to determine which cluster belongs to which tracked object the cluster centers, which are obtained by averaging the position of all scan points in the cluster, are compared with the positions of currently tracked objects, and each cluster is assigned to the nearest object. The clusters that are far from all currently tracked objects are considered as new objects, and a new tracking process is started for them.

3) Estimation of the object center and type

After the cluster centers are obtained and associated with objects the position of the tracked objects can be calculated. However, since each sensor can only make scan of only one side of an object, the obtained cluster center in general does not coincide with the center of the tracked object. In order to obtain the position of a tracked object we use the following relation, which is based on the assumption that the object has a circular shape (see Fig. 3):

$$(1) \qquad\qquad r_{obj} = r_{cl} + d_r,$$

where r_{cl} is the distance from the sensors to the cluster center and d_r is a parameter dependent on the dimensions (diameter) of the object. The angle α of the line between the laser range finder and the object is assumed to be the same as for the cluster center.

The approximation of the object shape with a circle can be considered appropriate since the objects we want to track - mobile robots and humans, i.e. human's legs - have a more or less a circular shape. The error introduced with the approximation is small and it is further reduced when multiple sensors track the object. Another source of error is the parameter d_r, which stands for the distance between cluster and object center. Although in general this distance is varying due to measurement noise, etc., it is approximated with

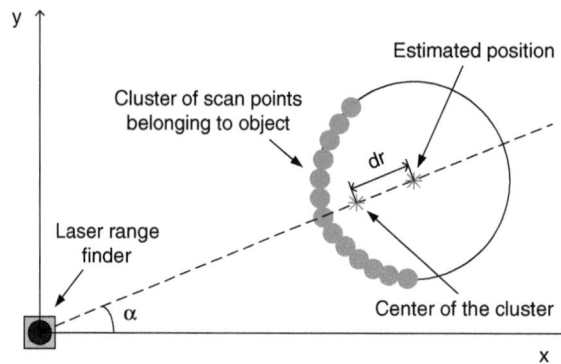

FIGURE 6. Tracking with laser range finders: the object center is assumed to be at a fixed distance from the center of the cluster of scan points belonging to it.

a constant value. In our experiments d_r was set to 6 cm for human legs and 15 cm for the mobile robot.

In the case of tracking robots, the position estimated using (1) can be considered as a measurement of the robot position. But in the case of tracking a human, the position has to be determined based on two clusters belonging to the human's legs, as noted earlier. Here we assume that the humans are not wearing a long skirt or similar clothes, in which case the legs would not be detected. In a given measurement step these clusters may or may not be available depending on occlusions in the scan. Since the position of the human can be assumed to be in the middle between both legs, in the case both legs are visible the measurement of the human's position is taken as their mean value. When only one leg is visible the other leg's position is predicted based on the current human's tracked position, and the measured position is calculated accordingly. However, in that case the measurement is considered less reliable.

The number of scan points in the cluster and the number of clusters belonging to an object can be used as a way to distinguish between humans and robots. The robot usually gets associated with one cluster, while tracked humans often get two clusters belonging to legs. We found out the type of object can be simply and robustly determined just by observing the averaged value of the number of associated clusters, and comparing it with a threshold.

4) Tracking using Kalman filter

The mobile robot we use is a differential drive robot. Its pose can be represented by the robot's (x, y) position and orientation θ in the space, which can be described by the following model:

$$
(2) \qquad \mathbf{x}(k+1) = \begin{bmatrix} x(k) + v(k)\Delta T \cos(\theta(k)) \\ y(k) + v(k)\Delta T \sin(\theta(k)) \\ \theta(k) + \Delta T \omega(k) \end{bmatrix} + \mathbf{V}\mathbf{v}(k),
$$

Here v is the translational velocity and ω is the rotational velocity as measured by the robot's wheel encoders. ΔT is the sampling time. The process noise $\mathbf{v}$ represents the uncertainty on the robot velocities, with $\mathbf{V}$ being the appropriate transformation matrix.

The noise is assumed zero mean Gaussian with covariance matrix $\mathbf{Q}$, which is equal to zero when the robot is stopped and grows linearly with increasing speed.

In the human tracking case the state is represented just by the (x, y) position and velocities v_x and v_y in the x and y direction. Here uniform motion is assumed since we have no specific knowledge of the object motion model. The human model is then given by the following linear state space model:

$$(3) \qquad \mathbf{x}(k+1) = \begin{bmatrix} x(k) + \Delta T v_x(k) \\ y(k) + \Delta T v_y(k) \\ v_x(k) \\ v_y(k) \end{bmatrix} + \mathbf{v}(k)$$

The noise $\mathbf{v}$ is assumed to be zero mean Gaussian, as in the robot case.

Using the procedure described in the previous section, laser range finders detect objects in the scan (humans or robot) and measure their position relative to the sensor. The outputs of the measurement are the range and angle to the detected objects, so the measurement model can be described as:

$$(4) \qquad \mathbf{z}(k) = \begin{bmatrix} q \\ \text{atan2}\,(\Delta x, \Delta y) + \theta_0 \end{bmatrix} + \mathbf{w}(k),$$

where Δx, Δy and q are given by the following relations:

$$(5) \qquad \Delta x = x_0 - x_1,$$

$$(6) \qquad \Delta y = y_0 - y_1,$$

$$(7) \qquad q^2 = \Delta x^2 + \Delta y^2.$$

The time step k was omitted to keep the notation simple. The indices 0 and 1 stand for the coordinates of the sensor (these are determined beforehand by calibrating the sensors, as described below in section 3) and observed object, respectfully. $\mathbf{w}$ represents the noise in the measurement, which is assumed zero mean Gaussian with covariance matrix $\mathbf{R}$. Apart from the actual noise in the scan, this noise also includes the error due to the approximation in equation (1) and the remaining uncertainty in the sensor pose after calibration.

Using the derived model and its linearization position tracking using an Extended Kalman Filter (EKF) can be implemented.

5) Fusion of multiple sensors

The process described above makes it possible for each laser range finder to independently track objects in the area it covers. But in order to implement tracking in a larger area and to achieve a more stable measurement by covering the same area with multiple sensors, estimates from multiple sensors are fused together. Every sensor sends the estimated positions and speeds to one central unit, called the fusion server (Fig. 3), where the estimates are combined to obtain positions of all tracked objects in the space.

A simple way to combine measurements is to take the mean of the estimated values as obtained on different sensors for each of the tracked objects. However, since the estimation is done using a Kalman filter the covariance of the estimate is also available, and it can be used in the fusion process. This results in the following equations for fusing estimates from n sensors [4]:

$$(8) \qquad \mathbf{P}_f = \left(\mathbf{P}_1^{-1} + \mathbf{P}_2^{-1} + \cdots + \mathbf{P}_n^{-1} \right)^{-1}$$

$$(9) \qquad \mathbf{x}_f = \mathbf{P}_f \left(\mathbf{P}_1^{-1}\mathbf{x}_1 + \mathbf{P}_2^{-1}\mathbf{x}_2 + \cdots + \mathbf{P}_n^{-1}\mathbf{x}_n \right)$$

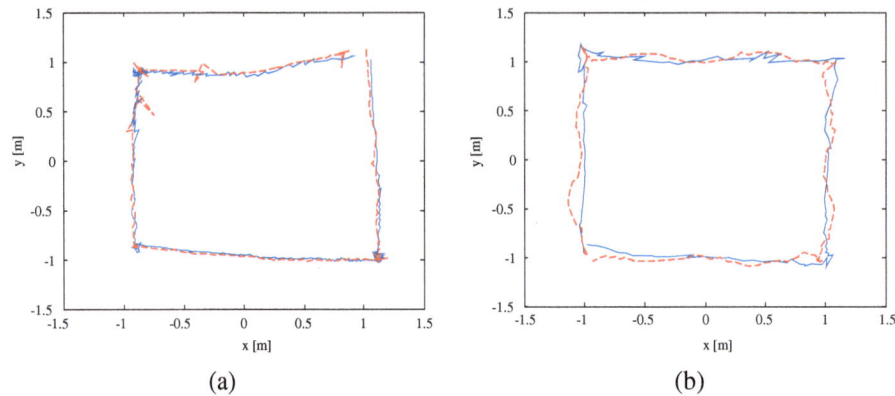

(a)   (b)

FIGURE 7. Tracking results - comparison of tracking results using laser range finders (solid line) with the ultrasound measurement (dashed line) (a) robot tracking (b) human tracking.

where $x_1, \cdots, x_n$ and $P_1, \cdots, P_n$, are the estimated values and the associated covariance matrices calculated on each sensor using the EKF equations. x_f and P_f are the obtained fused value and its covariance matrix.

6) Experiment - tracking the robot and humans

In order to test the tracking method we made several human and robot tracking experiments, where we used 4 laser range finders distributed in the space for tracking. An example of the obtained results is shown in Fig. 3 (a) and (b). In the experiment the robot and human were following a square shaped trajectory, although the robot's path deviated somewhat since it was driven manually.

Since there is no way to know the ground truth position of the tracked human or robot, we used the global positioning ultrasound system in our lab and compared it with the position obtained using laser range finders. With the ultrasound system the position is measured by putting two ultrasound transmitters on the robot, or in the case of humans by wearing a transmitter on the chest. In the robot tracking case this measurement is also combined with the output from the robot's wheel encoders. The ultrasound system gives a relatively accurate position measurement (for robot max. error around 10 cm) that can be considered as close to the real position of the tracked object.

It can be seen from the obtained robot tracking results (Fig. 3 (a)) that tracking using laser range finders gives very similar results to the one using the ultrasound system, and in fact the ultrasound system has more outliers in the estimate. In the human tracking case (Fig. 3 (b)) the difference between the two estimates is somewhat larger. Since in this case the ultrasound system measurements are directly used, the larger difference is partly due to the human swinging during walking – since the tag is attached to the human this swinging is reflected in the measurement, as can be noticed in the figure. However, the difference between the estimates is less than 10 cm for most of the time, which confirms the adequacy of the laser range finders based tracking system for application in iSpace.

Successful tracking of multiple objects was also obtained when the object density in the covered space did not cross 1 tracked object per square meter. In more crowded situations the system is still able to determine the position of all the objects, but possibility of errors in the object identity (i.e. finding out which object is which) increases.

Automated Calibration of Distributed Sensors. In order to be able to track the objects the sensors need to be calibrated, i.e. their position and orientation in space has to be known. Appropriate calibration is most important for data association and fusion of sensor measurements. Usually, calibration is done by using calibration objects that are objects with known position, shape and so on. The calibration procedure can be automated, for example by tracking mobile robots moving in the space and using these data for calibration. Some researchers focus on node localization in wireless sensor network using mobile robots [28,29]. In the methods, each mobile robot broadcasts its position information and if a wireless sensor node can get the information, the node is considered to be located adjacent to the robot. Effective path planning of mobile robots for wireless node localization is also addressed in [17]. Although the sensor nodes just receive the position information from the robots in these researches, the measurement of the sensors can also be used for calibration.

Fig. 3 shows the overview of the calibration method in case that a mobile robot in iSpace is utilized. Here we consider the automated calibration of the pose of the laser range finders. Our goal is to find transformation parameters (translation vector T and rotation matrix R) from the laser range finder coordinates to the world coordinates. Since laser range finders are usually placed horizontally in iSpace applications [10,34], the calibration parameters are position and orientation of the laser range finder in 2D plane $(T_{xg}, T_{yg}, \theta_g)$. Let $O - {}^W x^W y$ be the coordinate system fixed to iSpace (world coordinate system) and $C - {}^L x^L y$ be the coordinate system fixed to the laser range finder (laser range finder coordinate system). First, the DIND requests the mobile robot to send the position of the robot in world coordinate system (x_{k1}, y_{k1}). In a real environment a global positioning system like a ultrasound positioning system is usually not available, but it can still be possible to estimate the robot's position by implementing a self localization method based on a preexisting map, or by doing SLAM [31]. Each DIND also tracks the mobile robots and gets the position information in local coordinate system (x_{k2}, y_{k2}). The calibration process is then performed based on the set of corresponding points $\{(x_{k1}, y_{k1}), (x_{k2}, y_{k2})\}$ $(k = 1, 2, ..., n)$. This can be formulated as a least square error problem, whose solution gives the following calibration equations:

$$(10) \qquad T_{xg} = \mu_{x1} - \cos\theta_g \mu_{x2} + \sin\theta_g \mu_{y2}$$

$$(11) \qquad T_{yg} = \mu_{y1} - \sin\theta_g \mu_{x2} - \cos\theta_g \mu_{y2}$$

$$(12) \qquad \theta_g = \operatorname{atan2}\left\{\sum_{i=1}^{n} \frac{-(x_{i1}y_{i2} - y_{i1}x_{i2})}{n} + \mu_{x1}\mu_{y2} - \mu_{y1}\mu_{x2}, \right.$$
$$\left. \sum_{i=1}^{n} \frac{(x_{i1}x_{i2} + y_{i1}y_{i2})}{n} - \mu_{x1}\mu_{x2} - \mu_{y1}\mu_{y2}\right\}$$

where μ's stand for mean values, for example:

$$(13) \qquad \mu_{x1} = \frac{1}{n}\sum_{i=1}^{n} x_{i1}$$

The problem with least-squares estimation as given by equations (10)-(12) is sensitivity to outliers. Since robot tracking is done online it is possible that outliers, such as mis-correspondence between data can easily appear. In order to eliminate the effect of

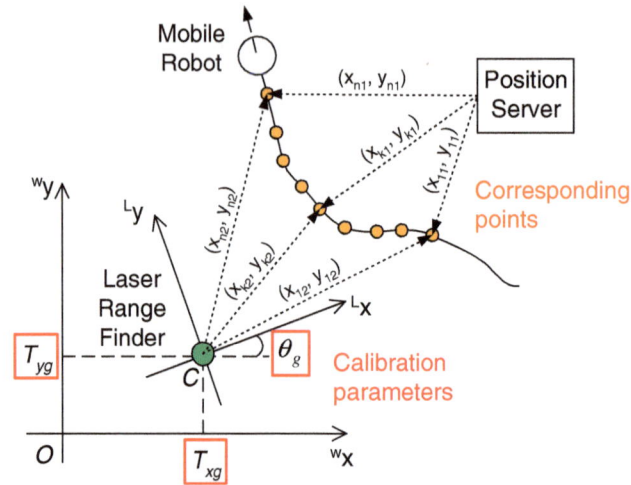

FIGURE 8. Overview of the calibration method.

TABLE 1. Comparison of automated and manual calibration results

Laser	Automated			Manual		
	T_{xg}[m]	T_{yg}[m]	θ_g[rad]	T_{xg}[m]	T_{yg}[m]	θ_g[rad]
1	0.99	0.95	-2.60	0.95	1.02	-2.54
2	-1.91	1.10	-0.19	-1.95	1.00	-0.15
3	0.09	-2.01	1.56	0.20	-1.97	1.60

outliers instead of simple least squares we use the least median of squares (LMedS) based estimation.

Therefore, the calibration process is summarized as follows:

(1) Store corresponding points (x_{k1}, y_{k1}), (x_{k2}, y_{k2}) acquired by the robot tracking process (as before 1 and 2 stands for global and local coordinates)
(2) Sample 2 data randomly from the set of corresponding points
(3) Calculate $(T_{xg}, T_{yg}, \theta_g)$ from the sampled data using equations (10)-(12)
(4) Evaluate the estimation error by the median of the square error for all corresponding points
(5) Repeat steps 2) – 4) N times
(6) Select $(T_{xg}, T_{yg}, \theta_g)$ which has minimum estimation error as the estimate

In the environment shown in Fig. 3, three laser range finders are calibrated using a mobile robot. The arrangement of the laser range finders and the path of the mobile robot estimated by the EKF are also shown. In the experiment the mobile robot entered the room from the right and circled around the room in clockwise direction.

Table 1 shows the result of the automated calibration compared to that of manual one. It can be seen that the result obtained using automated calibration is almost the same as when manual calibration is used.

The tracked position of the mobile robot in each sensor node, after transformation to global coordinates using the estimated parameters is shown in Fig. 3. It is obvious that there

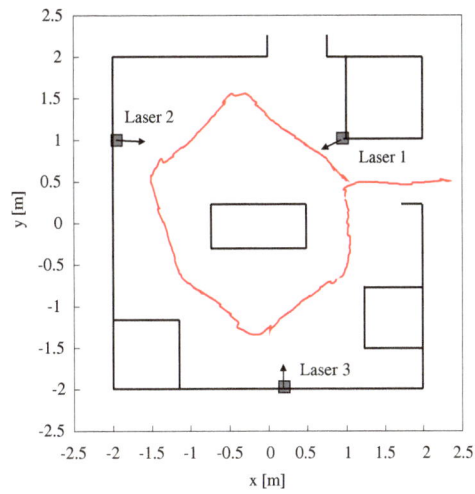

FIGURE 9. Experimental environment – arrangement of the laser range finders and path of the mobile robot estimated by the extended Kalman filter.

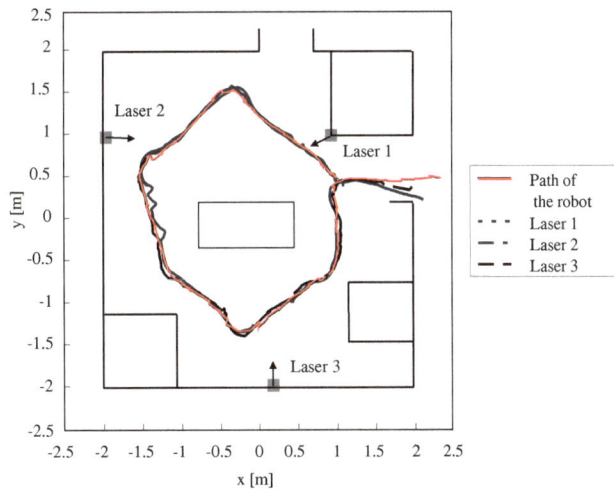

FIGURE 10. The actual path of the mobile robot (presented again) and the tracked positions of the robot in each sensor node, obtained by transformation into global coordinates using the estimated parameters.

is a good correspondence between the tracks, which shows that the calibrated parameters are correct.

Although calibration of a laser range finder is considered in this section, a similar approach can be applied to other sensors including cameras [23, 26]. For example, camera calibration can be performed based on the positions of the robot in the world coordinate system and their corresponding points in image coordinate system by applying Tsai's method [32].

4. Recognition Function

Spatial-Human Interface - Spatial Memory. iSpace has to recognize requests from users to provide desired services. But it is desirable that the user can activate the services through natural interfaces. Therefore, a suitable human-iSpace interface is needed. To realize this, the spatial memory system has been proposed as an interface between users and iSpace [20]. In this system, we adopt indication actions of users as operation methods in order to achieve an intuitive and instantaneous way that anyone can apply. A position on user's body is called a human indicator. When a user specifies digital information and indicates a position in the space, the system associates the three-dimensional position with the information and manages the information as Spatial-Knowledge-Tag (SKT). Therefore, users can store and arrange computerized information such as digital files, robot commands, voice messages etc. into the real world. They can also retrieve the stored information in the same way as on storing action, i.e. indicating action. Thus, we named the system "spatial memory."

While working in the real world, users can obtain environmental information such as arrangements of equipment including desks, file cabinets and so on. Consequently, the users can arrange digital information in a much easier way and memorize the whereabouts of the information by referring to such environmental information. Also their spatial cognition capabilities such as the motion sense, are utilized for memorizing the whereabouts, and will prompt the users to recall them.

We carried out some experiments to confirm the usefulness of the system.

First, the usability of the spatial memory was investigated from the viewpoints of the accessibility and the effectiveness of memorizing. For these investigations, the task of learning the contents and the whereabouts of stored seven SKTs and accessing a randomly specified SKT was carried out several times by changing intervals of tasks. Time variation of the task completion time was obtained. The results showed that the completion time of all subjects became much shorter while increasing the number of performances, although there were large differences among the completion time of the subjects. In addition, all subjects successfully completed getting access to all SKTs in the performance even after about four weeks from the first performance and the last performance after 20 days duration was as short as the performance of two-hour duration. These results confirmed that the accessibility was improved and the spatial memory improved memorizing the stored computerized information once the spatial arrangement was learned.

The second experiment was carried out in order to check the usefulness of the spatial memory by comparing the two cases: (A) using the spatial memory and (B) not using the spatial memory (using a conventional computer). The subjects carried out a specified task which is to input items into text files after finding out specified items from given contents by using the spatial memory and only using a conventional computer. To evaluate the usefulness, the performance efficiency was quantified as the ratio of the input work time to the total performance time, and the averaged performance efficiencies of the two cases were compared. The results confirmed that the time required for the additional processes such as access and saving computerized information was reduced by using the spatial memory.

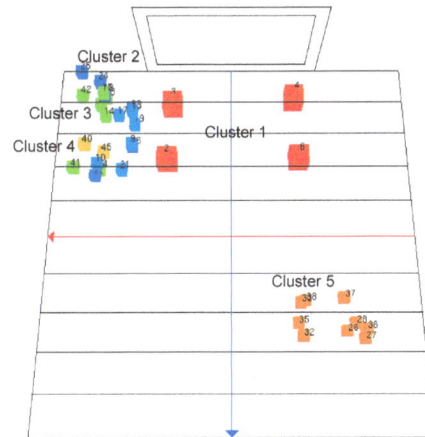

FIGURE 11. Clustered SKTs that correspond each human activity such as "Listening to music," "Writing a paper," "Watching movies," etc.

Applying the t-test for two correlated samples under the 5% level of significance, significant difference of the input efficiency in the two performances was confirmed.

Recognition of Human Activity. Although users can utilize Spatial Memory to send the requests to iSpace, interaction can also be started by the space. If iSpace finds that a user is in trouble based on the observation, for example, a mobile robot in the space would go to help the user. In order for iSpace to provide various services for people in a wide range of situations, current intended purposes of the activities should be recognized.

The users actively create SKTs according to their needs and purposes of their activities. As a result, arranged SKTs in an environment correspond to activity histories of the users. Therefore, we can consider that classification of arranged SKTs leads to classification of human activities and estimation of intended purposes for places. We performed a classification of arranged SKT based on observation of human activities and forms of spatial memory data. Usage histories of the spatial memory can be regarded as abstraction of human activities. We use unsupervised hierarchical clustering based on SKT similarity defined by using the usage history.

Fig. 4 shows an example of the result of clustering. Here distance of two SKTs are defined by the linear combination of the physical distance between the SKTs, difference of the size (access region) of the SKTs and the difference of the contents of the SKTs. The observation is done for about two days. The result confirmed that the approach was able to classify human activities not only spatially but also characteristically.

However, the problem of the approach is that human activities using the spatial memory are only observable. We usually utilize physical objects to accomplish our activities. Therefore, we consider that physical objects are also necessary information to describe human activities more precisely, especially in the case when we focus on objects used by people to accomplish their activities.

We define that object information represents the property of an object. The property of an object can be described based on two kinds of information: static information and non-static information. Information about an object's name, its color and size can be given because we can consider that these kinds of information will not change. Thus, such information can be regarded as static information. On the other hand, information including frequency of use, location, users and motion patterns cannot be determined in advance, because they depend on individuals or contexts when objects are used. These kinds of non-static information are important to describe human activities. Therefore, we decided to observe physical interactions between objects and users, and we attempt to obtain non-static object information from the observation of human-object relations.

Based on the observation, we describe non-static object information by using "who," "what," "when," "where," and "how (motion patterns)," (4W1H) information [21]. In order to detect the occurrence of physical interaction, a sensor node which consists of an acceleration sensor and a wireless network transmitter is attached on each object and users' hand. An ultrasound transmitter and a motion sensor are also attached to each human hand. "Who" and "what" information can be obtained by the IDs of the sensor nodes and "when" information is determined by time stamp information of the motion data. "When" and "How" information is acquired by data from ultrasound transmitter and motion sensor, respectively. We are currently working to describe human activities based on 4W1H information.

Extraction of Path Patterns from Human Walking Paths. The information about paths frequently used by human is important for mobile robots in order to operate with minimal disturbance to humans. For example, the information of path patterns can be used for paths of mobile robots since the area where human walks is also traversable for the mobile robots [3, 30]. In addition, by predicting the future motion of currently tracked human from the history of the observed paths in the environment, mobile robots would avoid unnecessary contact with people [5, 24]. Detection of unusual paths are also important for security or guidance applications in order to determine suspicious people or people who get lost.

So, we extract frequently used paths from the obtained walking paths. This is done in three steps: 1) acquisition of human walking path, 2) extraction of important points and 3) path clustering and averaging.

Since acquisition of human walking path is discussed in section 3, we first explain the extraction of important points. Here we define important points as entry/exit points, which are useful for mobile robots to move from one area to another, and stop points, which are helpful when mobile robots approach humans to provide services. The entry/exit points are extracted based on the points where the tracking system finds new objects or loses objects. On the other hand, the stop points are extracted based on the points where the speed of a human is lower than a threshold for some number of consecutive time steps. These candidates for entry/exit and stop points are grouped by hierarchical clustering and considered as important points if a cluster which consists of many points is formed. We use Euclidean distance in the $x - y$ plane as measure of distance between the points. The clustering process is continued until the distance between clusters exceeds a certain value because it is hard to determine how many important points are in the environment.

In the next step, for all combinations of two important points, we consider paths, which have these points for start and goal points. If there is more than one path that connects the two points path clustering is performed. We use a hierarchical clustering method based on the LCSS (Longest Common Subsequence) similarity measure (S1 similarity function presented in [33]). The LCSS models measure the similarity between two trajectories A and B based on how many corresponding points are found in A and B. This model allows time stretching so the points which have close spatial position and the order in the path

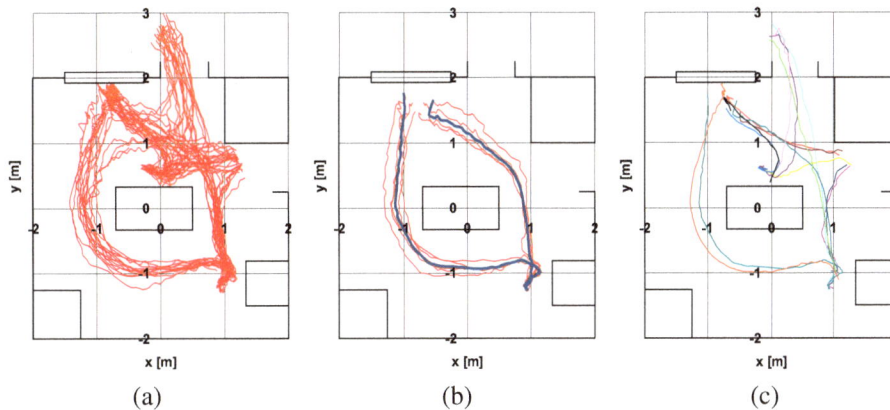

FIGURE 12. Snapshot of development support tools (a) obtained human served walking paths (b) examples of averaging (c) averaged waking paths.

can be matched. The best match obtained under the condition that the rearranging of the order of the points is prohibited is used for calculation of the similarity. The ratio of the number of corresponding point to the number of points in the shorter path is defined as the similarity. Finally, clustered paths are averaged to extract frequently used paths in the environment. An averaged trajectory is derived from corresponding points between two trajectories, which can be obtained from the LCSS similarity measure. The middle point of corresponding points is used to acquire averaged paths.

An example of human walking paths obtained by the tracking system is shown in Fig. 4 (a). Fig. 4 (b) shows an example of averaging for the path between two important points Nine walking paths are acquired and two paths are obtained by averaging. If the paths were calculated using the mean value, the obtained path would be located in the center of them, which means it would collide with a desk. The proposed method distinguishes the whole paths by using path clustering so that it is able to average clustered paths independently. Fig. 4 (c) shows all the averaged paths. We can observe that two paths were obtained in almost the same location between almost every important point. Pairs of paths were obtained because the direction in which the humans walked was taken into account. By including the direction information, various details about the environment can be extracted, e.g. the robot should keep to the right side here, this street is one-way, etc.

5. Actuation Function

Mobile Robot Navigation. This subsection focuses on control a mobile robot and move it through the space toward a goal. Although this by itself might not be a very useful action, it represents the basis for different physical services, such as object carrying, human guiding, etc.

By using the map and position of humans and robot in the space obtained by iSpace, it is possible to implement the control of the actuator inside the iSpace. The control process consists of path planning and path tracking and obstacle avoidance. Even if the robot does not have any intelligence implemented on it, iSpace can perform the calculations and send the speed commands to the robot.

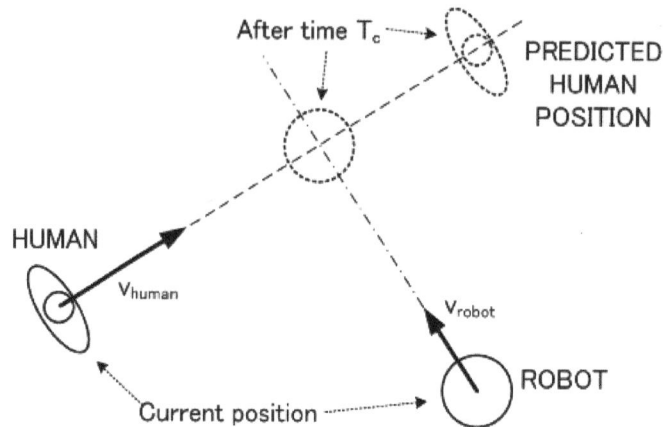

FIGURE 13. Prediction of human position used in path planning.

1) Path planning

Path planning is the problem of finding the best way for the robot to move through the space based on the map of the obstacles in the space. As usual in the robot control, the map obtained during the mapping process is represented as an occupancy grid. That is why the most often used path planning methods are various graph search methods. In our implementation we use the Field D* method [9], which gives arbitrary path headings and is not restricted to multiples of 45° as other existing path planning methods. The main idea is to calculate the cost of the cell corners, and to linearly interpolate the cost on the cell edges. Although this linear interpolation approach is not exact, it gives a good approximation of the real cost, and it is possible to calculate the cost of both the points on the edge and inside the cell.

Before planning, the obstacles in the map, both static and moving, are enlarged by the radius of the robot. Also, in order to keep the robot from moving too close to the obstacles, in the vicinity of obstacles the cost of traversing the cells is higher (the closer to the obstacle the higher the cost). The Field D* method is applied on the obtained grid, which gives a smooth path from the robot to the goal.

The Field D* and other planning methods treat all the obstacles as static. But because the existence of moving obstacles, like humans in the case of iSpace, the obtained path might not be optimal since the moving obstacles might affect it in future. In order to achieve improved planning we utilize the estimates of the robot and humans speed obtained in the tracking process. Based on the robot's and human's current speeds, we estimate the position the human will have in the moment the robot crosses his/her path, as shown on Fig. 5. Then this predicted human position is used instead of the real position in the path planning algorithm. The predicted position is only an approximation, but it becomes better the closer the robot is to the human. This is a very simple method but it gives an improved path planning in presence of moving obstacles, as proven in experiments.

2) Path tracking and obstacle avoidance

In order for the robot to follow the calculated path and at the same time avoid bumping into obstacles, a local control algorithm is used. A variety of approaches for tracking control of nonholonomic mobile robots exist for example, the control law which is determined by using Lyapunov function [**15**], the control method based on the dynamic feedback linearization [**22**], etc. In our implementation we use the Dynamic Window Approach (DWA) [**11**], because of its physically meaningful representations. The DWA is used to generate the mobile robot speed commands that ensure that the robot does not collide with obstacles and that it follows the desired path. Moreover, the obtained commands are calculated in such way that they do not violate the dynamic constraints of the robot, i.e. its maximal transversal and rotational accelerations. In order to achieve that, calculations are done directly in the velocity space of the robot, that is, the space defined by its speed and angular speed. First, the combinations of transversal and angular speed can be reached in the next step based on the allowed maximal accelerations are determined. Then an objective function that includes obstacle avoidance, speed and goal heading criteria is used to determine which speed combination is most appropriate to be applied in the next time step.

3) Experiment

We tested the developed system in our experimental space. Three laser range finders were arranged in the space and using them a map of the space was built and humans and a mobile robot were continuously tracked. Based on the obtained map and positions the mobile robot was driven in the environment by the iSpace.

Fig. 5 shows an example of the obtained robot navigation results. In this situation the iSpace was given a command to drive a robot to the point with coordinates (1, 1). At that time a human was also present in the space. Based on the robot speed and the estimated human speed (illustrated as arrows) the position of the human is predicted as described earlier (shown as stars) and used in the path planning. As a result the iSpace smoothly guided the robot behind the human and toward the goal.

Information Display Using Pan-Tilt Projector. The information display system uses a projector with a pan-tilt unit, which is able to project an image toward any position to realize interactive information display according to human movement in the space. By utilizing the interactive information, many applications can be developed, for example, the display of signs or marks in public spaces, or various information services in daily life. However, main issues in active projection are compensation of projection image and occlusion avoidance.

1) Compensation of projection image

When projection direction is not orthogonal to the projection surface, projection distortion occurs. Moreover the size of the projected image depends on the distance to the projection surface. Therefore with the change of the projection point it is not possible to provide a uniform image to a user. To provide a uniform projection toward any position the projector implements a compensation of the projection image by using a geometric model and inverse perspective conversion.

Projector light radiates out so that the projected image size depends on the distance to the projection surface. By calculating the relation of those factors, it is able to pre-compensate the projection image size to the desired size W. Here the resize ratio γ is given as follows.

$$\gamma(d) = W/t(d) \tag{14}$$

where d denotes distance between the projector and the projection surface and $t(d)$ is the image size on the projection surface.

Distortion is also caused by the angle between the optical axis of the projector and the projection surface. The geometrical definition is shown in Fig. 5.

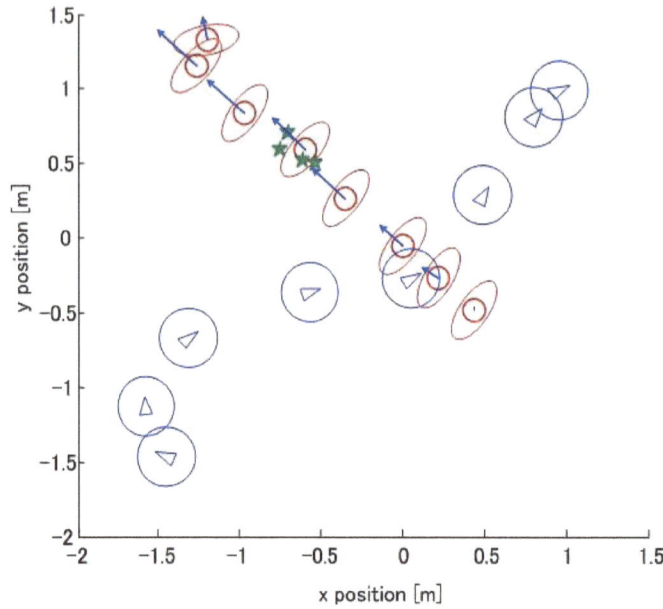

FIGURE 14. Guiding the robot. The human's future positions are estimated and used in path planning calculations, so the robot is guided around this predicted obstacle from the beginning.

As shown in this figure, the pan-tilt projector projects an image toward O_p. The plane Q is the projection surface and the plane R is orthogonal to the projection direction. The points r_1 to r_4 denote the corners of the non-distorted image whereas the points q_1 to q_4 are the corresponding points on the distorted image. A relation between a point p_Q on plane Q and a point p_R on plane R is obtained based on perspective conversion.

$$(15) \qquad \begin{bmatrix} p_Q \\ 1 \end{bmatrix} \sim H_{QR} \begin{bmatrix} p_R \\ 1 \end{bmatrix}$$

This conversion matrix H_{QR} is a 3×3 matrix and the degree of freedom is 8. Therefore, if four or more sets of corresponding points of p_Q and p_R are given, we can identify H_{QR} and represent image distortion. The corresponding points can be found by the intersection of the plane Q with the line through r_i from the projection origin (lens). The inverse matrix of H_{QR} represents compensation of image distortion and we can get the pre-compensated output image. We use a normal vector of the plane Q for finding the corresponding points. The plane Q shows the projection surface so that it is possible to compensate projection toward any surface by setting normal vector according to the space.

2) Occlusion avoidance

Projection occlusion occurs when human enters into the area where the human obstructs the projection. This problem sometimes happens in active projection due to human movement or change in the environment. Hence, by creating an occlusion area and a human (obstruction) model and judging whether they overlap with each other, occlusion can be detected and avoided.

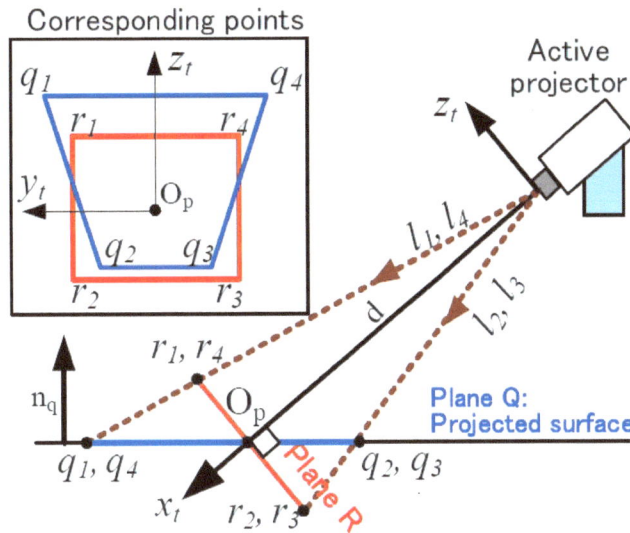

FIGURE 15. Geometrical definition of image distortion.

We modelled the shape of projection light and human are corn and cylinder, respectively. We judge the overlap between these two models to detect occlusion. Moreover, not only human but also other objects including chairs and tables could cause the occlusion problem. Our occlusion avoidance algorithm can also be used for those objects by considering their shape models.

The avoidance method needs to modify the projected position so that the user can easily view the image. Fig. 5 shows the determination of the modified position. In the situation that the projection position is on the left side of the human model, the projection direction is moved to the left to avoid occlusion since it requires less angular variation compared to the rightward movement. On the contrary, when the projection position is on the right side of the human model, it moves to the right for the same reason. If the calculated correction angle is greater than the limit correction angle θ_{max}, the projection position is moved away from the human.

3) Experiment

In this experiment, the information display system performs projection toward the front of a user. Fig. 5 shows the results of the projection for various poses of the user. The system performs the projection of double-rectangle. As shown in Fig. 5 (a), the projection without compensation generates distorted image. On the other hand, the projection with compensation provides correct information (considered size, direction and distortion of projection image). Fig. 5 (b) shows the results of several situation in the experimental space. When occlusion occurs (two cases), the projection positions are modified.

6. Conclusion

Intelligent Space (iSpace) is a platform on which it is possible to implement advanced technologies that enable easy realization of smart services to human. In this chapter, our ongoing research on Intelligent Space Project and achieved results are presented.

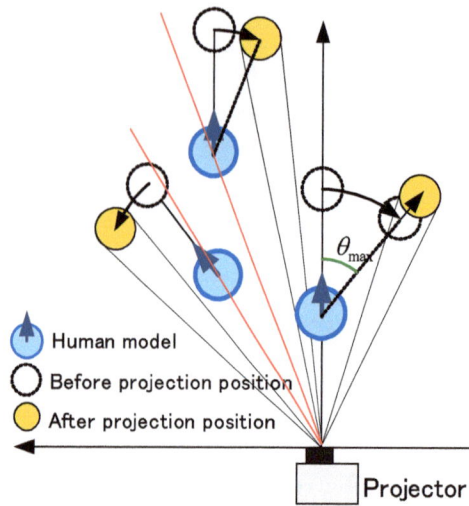

FIGURE 16. Determination of the modified position.

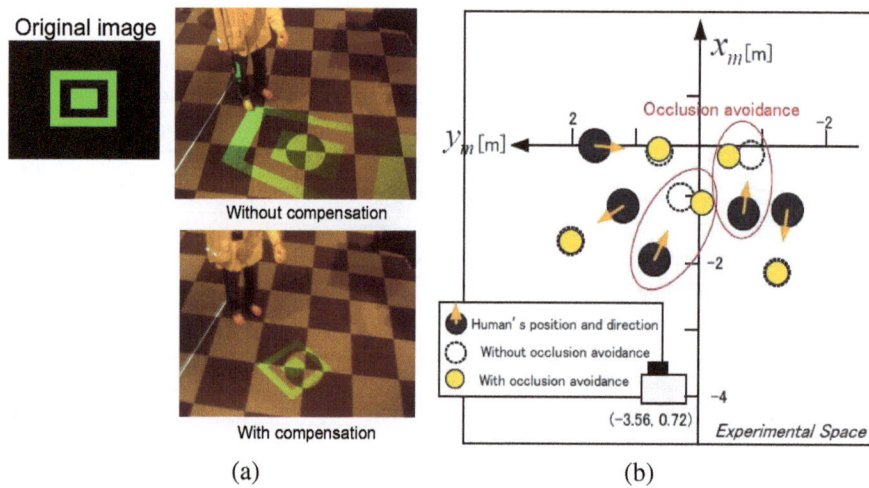

FIGURE 17. Results of projection using a pan-tilt projector (a) distortion compensation (b) projected positions.

Intelligent robot systems are developed by integration of mechatronics and software technologies. However, the systems are getting more complicated since the cooperation of various types of robots is necessary to realize advanced services for users as also shown in the network robot [12] and the ubiquitous robot [16] concept. Therefore, the system integration becomes an important issue. First we discussed design and implementation of iSpace system In order to realize a flexible and scalable system, iSpace is implemented using RT (robot technology) middleware.

Moreover, in order to support human in informative and physical ways, iSpace has three key functions: "observation," "recognition," and "actuation." We introduced the researches on each function. The "observation function" of iSpace is developed using the distributed sensory network. The space is able to track multiple humans and robots, and also handle the case of occlusion between objects by using multiple sensors. The observed information is also useful for the development of the system. In the case of iSpace, it takes a great deal of time and effort to calibrate the sensors since multiple sensors are distributed in the space. By utilizing the observed results, the calibration procedure can be automated. We also described our approach to extract useful information regarding the environment using the observations of the space, such as frequently used walking paths and human activities. The information is used by the iSpace to plan its actions. The spatial memory is also an efficient technology to handle the information exchange between human and iSpace, and gives the space the ability to recognize and track human activities. As the "actuation function" of iSpace, we presented the implementation of mobile robot navigation and the information display system using a pan-tilt projector.

Bibliography

[1] Akiyama, T., J.-H. Lee and H. Hashimoto, "Evaluation of CCD Camera Arrangement for Positioning System in Intelligent Space," Proc. of Seventh Int. Symp. Artificial Life and Robotics, pp.310–315, (2002).

[2] Ando, N., T. Suehiro, K. Kitagaki, T. Kotoku and W.-K. Yoon, "RT-Middleware: Distributed Component Middleware for RT (Robot Technology)," Proc. of IEEE/RSJ Int. Conf. on Intelligent Robots and Systems, pp.3555–3560, (2005).

[3] Appenzeller, G., J.-H. Lee and H. Hashimoto, "Building Topological Maps by Looking at People: An Example of Cooperation between Intelligent Spaces and Robots," Proc. of the 1997 IEEE/RSJ Int. Conf. on Intelligent Robots and Systems, 3:1326–1333, (1997).

[4] Bar-Shalom, Y. and T. E. Fortmann, Tracking and Data Association, Academic Press, (1988).

[5] Bennewitz, M., W. Burgard, G. Cielniak and S. Thrun, "Learning Motion Patterns of People for Compliant Robot Motion," The Int. Jnl. of Robotics Research, 241:31–48, (2005).

[6] Brooks, A., T. Kaupp, A. Makarenko, S. Williams and A. Oreback, "Orca: a Component Model and Repository," Software Engineering for Experimental Robotics (edited by D. Brugali), Springer Tracts in Advanced Robotics, 30:231–251, (2007).

[7] Broxvall, M. "A Middleware for Ecologies of Robotic Devices," Proc. of the First Int. Conf. on Robot Communication and Coordination, 30 (1)–(8), (2007).

[8] Cook, D. J. and S. K. Das, Smart Environments: Technologies, Protocols, and Applications (Wiley Series on Parallel and Distributed Computing), Wiley-Interscience (2004).

[9] Ferguson, D. and A. Stentz, "The Field D* Algorithm for Improved Path Planning and Replanning in Uniform and Non-uniform Cost Environments," Technical Report CMU-RI-TR-05-19, Robotics Institute, (2005).

[10] Fod, A., A. Howard and M. J. Mataric, "A Laser-Based People Tracker," Proc. of the 2002 IEEE Int. Conf. on Robotics and Automation, 3:3024–3029, 2002.

[11] Fox, D., W. Burgard and S. Thrun, "The Dynamic Window Approach to Collision Avoidance," IEEE Robotics and Automation Magazine, 41:23–33, (1997).

[12] Hagita, N., "Communication Robots in the Network Robot Framework," Proc. of the 9th Int. Conf. on Control, Automation, Robotics and Vision, pp.1–6, (2006).

[13] Hightower, J. and G. Borriello, "Location Systems for Ubiquitous Computing," IEEE Computer, 348:57–66, (2001).

[14] Johanson, B., A. Fox and T. Winograd, "The Interactive Workspaces project: Experiences with Ubiquitous Computing Rooms," IEEE Pervasive Computing, 12:67–74, (2002).

[15] Kanayama, Y., Y. Kimura, F. Miyazaki and T. Noguchi, "A Stable Tracking Control Method for a Non-Holonomic Mobile Robot," Proc. of the IEEE/RSJ Int. Workshop on Intelligent Robots and Systems, 3:1236–1241, (1991).

[16] Kim, J.-H., K.-H. Lee, Y.-D. Kim, N. S. Kuppuswamy and J. Jo, "Ubiquitous Robot: a New Paradigm for Integrated Services," Proc. of the 2007 IEEE Int. Conf. on Robotics and Automation, pp.2853–2858, (2007).

[17] Koutsonikolas, D., S. M. Das and Y. C. Hu, "Path Planning of Mobile Landmarks for Localization in Wireless Sensor Networks," Proc. of the 26th IEEE Int. Conf. on Distributed Computing Systems Workshops, 86 (1)–(7), (2006).

[18] Krumm, J., S. Harris, B. Meyers, B. Brumitt, M. Hale and S. Shafer, "Multi-Camera Multi-Person Tracking for EasyLiving," Proc. of the 3rd IEEE Int. Workshop on Visual Surveillance, pp.3–10, (2000).

[19] Morioka, K. and H. Hashimoto, "Appearance Based Object Identification for Distributed Vision Sensors in Intelligent Space," Proc. of 2004 IEEE/RSJ Int. Conf. on Intelligent Robots and Systems, pp.199–204, (2004).

[20] Niitsuma, M., H. Hashimoto and H. Hashimoto, "Spatial Memory as an Aid System for Human Activity in Intelligent Space," IEEE Trans. on Industrial Electronics, 542:1122–1131, (2007).

[21] Niitsuma, M., K. Yokoi and H. Hashimoto, "Describing Human-Object Interaction in Intelligent Space," Proc. of the 2nd Int. Conf. on Human System Interaction, pp.395–399, (2009).

[22] Oriolo, G., A. De Luca and M. Vendittelli, "WMR Control via Dynamic Feedback Linearization: Design, Implementation, and Experimental Validation," IEEE Trans. on Control Systems Technology, **106**:835–852, (2002).

[23] Rekleitis, I. and G. Dudek, "Automated Calibration of a Camera Sensor Network," Proc. of IEEE/RSJ Int. Conf. on Intelligent Robots and Systems, pp.3384-3389, (2005).

[24] Rennekamp, T., K. Homeier and T. Kroger, "Distributed Sensing and Prediction of Obstacle Motions for Mobile Robot Motion Planning," Proc. of the 2006 IEEE Int. Conf. on Intelligent Robots and Systems, pp.4833–4838, (2006).

[25] Rudolph, L., "Project Oxygen: Pervasive, Human-Centric Computing - an Initial Experience," Proc. of the 13th Int. Conf. on Advanced Information Systems Engineering, pp.1–12, (2001).

[26] Sasaki, T. and H. Hashimoto, "Camera Calibration using Mobile Robot in Intelligent Space," Proc. of SICE-ICASE Int. Joint Conf. 2006, pp.2657–2662, (2006).

[27] Sasaki, T. and H. Hashimoto, "Component Based Integration of Intelligent Space and its Application to Mobile Robot Navigation," Proc. of the 2009 IEEE/ASME Int. Conf. on Advanced Intelligent Mechatronics, pp.1486–1491, (2009).

[28] Shenoy, S. and J. Tan, "Simultaneous Localization and Mobile Robot Navigation in a Hybrid Sensor Network," Proc. of IEEE/RSJ Int. Conf. on Intelligent Robots and Systems, pp.1636–1641, (2005).

[29] Sreenath, K., F. L. Lewis and D. O. Popa, "Localization of a Wireless Sensor Network with Unattended Ground Sensors and Some Mobile Robots," Proc. of the 2006 IEEE Conf. on Robotics, Automation and Mechatronics, pp.1–8, (2006).

[30] Tanaka, K., N. Okada and E. Kondo, "Building a Floor Map by Combining Stereo Vision and Visual Tracking of Persons," Proc. of the 2003 IEEE Int. Symp. on Computational Intelligence in Robotics and Automation, **2**:641–646, (2003).

[31] Thrun, S., W. Burgard and D. Fox, Probabilistic Robotics, MIT Press, (2005).

[32] Tsai, R. Y., "A Versatile Camera Calibration Technique for High-Accuracy 3D Machine Vision Metrology Using Off-the-Shelf TV Cameras and Lenses," IEEE Jnl. of Robotics and Automation, **RA-3**4:323–344, (1987).

[33] Vlachos M., G. Kollios and D. Gunopulos, "Discovering Similar Multidimensional Trajectories," Proc. of 18th Int. Conf. on Data Engineering, pp.673–684, (2002).

[34] Zhao, H. and R. Shibasaki, "A Novel System for Tracking Pedestrians Using Multiple Single-Row Laser-Range Scanners," IEEE Trans. on Systems, Man and Cybernetics - Part A: Systems and Humans, **35**2:283–291, (2005).

[35] Zheng, S., M. Niitsuma, T. Sasaki and H. Hashimoto, "Research on Intelligent Space Architecture Developing Tools," Proc. of the 5th Int. Conf. on Ubiquitous Robots and Ambient Intelligence, pp.504–509, (2008).

Send Orders of Reprints at reprints@benthamscience.net

CHAPTER 8

Universal Design of Ubiquitous Robotic Space

Bong Keun Kim, Hyun Min Do, Hideyiki Tanaka, Yasushi Sumi, Hiromu Onda, Tamio Tanikawa, and Kohtaro Ohba

Intelligent Systems Research Institute

National Institute of Advanced Industrial Science and Technology (AIST)

Tsukuba, Japan

and

Tetsuo Tomizawa

Graduate School of Information Systems

The University of Electro-Communications

Tokyo, Japan

ABSTRACT. In order to cope with many problems of everyday activities which requires robots to accomplish complex yet routine tasks in changing environments, this study focuses on universality and proposes seven principles of universal design with robots. These principles are used to design ubiquitous robotic spaces. Further, in order to manage and to control the ubiquitous robotic spaces with robots, ambient intelligent systems, and everyday objects, we define inherent functions of objects as ubiquitous functions, and propose a ubiquitous function service with service-oriented architecture for the integration and control of physical and virtual objects under a unified framework, which is composed of smart object service, smart logic service, and smart discovery service. In addition, in order to provide flexible robustness to a robot functioning in the everyday domain, a reproducible robust internal-loop compensator structure is proposed, and the structural design of smart logic service based on this structure is illustrated.

1. Introduction

Ubiquitous computing integrates robot technology into the surrounding environments and allows robots to share information and knowledge of distributed objects through networks. Thus, robots emerge from the industrial domain to assist humans in their everyday lives. Ubiquitous network technology and knowledge-distribution systems facilitate this radical change [1]. Various technologies have been developed with this end in view [2–4], and the applications of distributed robotics are increasing in number. In particular, distributed control systems utilizing web technology enable interaction between heterogeneous systems [5,7]. Further, knowledge-sharing robot systems which associate distributed objects in the physical space with the knowledge DB available in the virtual space using RFID tags functioning physical hyperlinks, are good examples of robot systems based on ubiquitous computing [6,8,10].

Ubiquitous computing allows robots to perform highly complex tasks and leads to the diversification of the components of distributed objects. However, it is still difficult to optimize the robot control architecture designed for a specified goal in a ubiquitous robotic space which consists of multiple agents. In order to cope with this problem, this study emphasizes the universality of robot software and hardware systems, and introduces universal design with robots. Universal design is the design of products and environments to be usable by all people, to the greatest extent possible, without the need for adaptation or specialized design [9]. In this chapter, since users of ubiquitous robotic spaces and everyday objects in these spaces are not only human beings but also robots, it is necessary to redefine universal design considering robot characteristics. As a result, the universal design of ubiquitous robotic spaces enables every robot to use them easily independently on hardware and software structure for motion control and information exchange among agents. Fig. 1 shows the proposed seven principles of the universal design for ubiquitous robotic space. Based on this, a ubiquitous function (UF) service is proposed for ubiquitous robotic spaces.

(1) Equitable Use: The design is useful and applicable to robots with diverse mechanisms and abilities. The guidelines for this rule can be described as follows:
   (a) Provide the same means of use for all robots.
   (b) Avoid segregating or stigmatizing any robots.
   (c) Provisions for privacy, security, and safety should be equally available to all robots.
   (d) Make the design appealing to all robots.

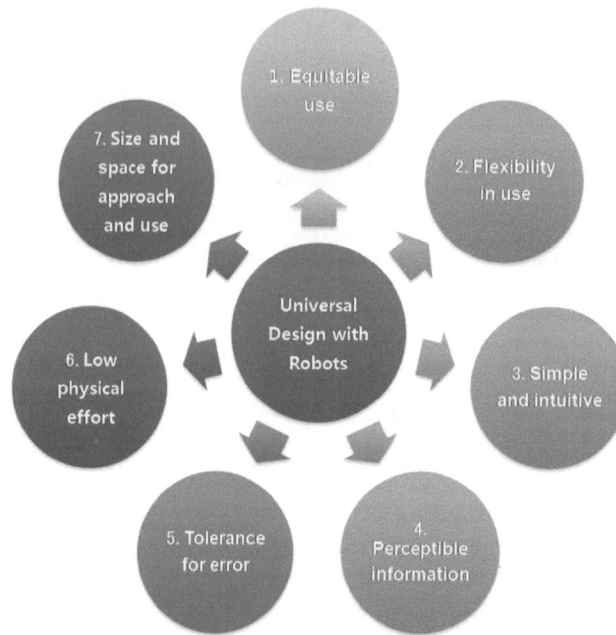

FIGURE 1. Seven principles of universal design with robots.

(2) Flexibility in Use: The design accommodates a wide range of robots' preferences and abilities. The guidelines for this rule is as follows:

(a) Provide choice in methods of use.

(b) Accommodate mechanism free access and use.

(c) Facilitate the robot's accuracy and precision.

(d) Provide adaptability to the robot's pace.

(3) Simple and Intuitive: Use of the design is easy to understand, regardless of the robot's sensing and recognition abilities. The guidelines for this rule is as follows:

(a) Eliminate unnecessary complexity.

(b) Be consistent with the robot's recognition ability.

(c) Accommodate a wide range of knowledge and information technologies.

(d) Arrange information consistent with its importance.

(e) Provide effective prompting and feedback during and after task completion.

(4) Perceptible Information: The design communicates necessary information effectively to the robot, regardless of ambient conditions or the robot's sensory abilities. The guidelines for this rule is as follows:

(a) Use different modes for redundant presentation of essential information.

(b) Provide adequate contrast between essential information and its surroundings.

(c) Maximize legibility of essential information.

(d) Differentiate elements in ways that can be described.

(e) Provide compatibility with a variety of techniques or devices used by robots with sensory limitations.

(5) Tolerance for Error: The design minimizes hazards and the adverse consequences of accidental or unintended actions. The guidelines for this rule is as follows:

    (a) Arrange elements to minimize hazards and errors: most used elements, most accessible; hazardous elements eliminated, isolated, or shielded.

    (b) Provide warnings of hazards and errors.

    (c) Provide fail safe features.

    (d) Discourage unmanageable action in tasks that require vigilance.

(6) Low Physical Effort: The design can be used efficiently and comfortably and with a minimum of fatigue. The guidelines for this rule is as follows:

    (a) Allow robot to maintain a neutral body position.

    (b) Use reasonable operating forces.

    (c) Minimize repetitive actions.

    (d) Minimize sustained physical effort

(7) Size and Space for Approach and Use: Appropriate size and space is provided for approach, reach, manipulation, and use regardless of robot's body size, posture, or mobility. The guidelines for this rule is as follows:

    (a) Provide a clear line of sight to important elements for any mobile or manipulator robot.

    (b) Make reach to all components comfortable for any mobile or manipulator robot.

    (c) Accommodate variations in robot hand and grip size.

    (d) Provide adequate space for the use of assistive devices or personal assistance.

2. Ubiquitous Function Services

Computer networks pervade the physical and virtual worlds, providing a variety of services in our lives. Network-based distributed control systems such as the knowledge-sharing robot systems associate physical objects with knowledge information available in the virtual space through networks and allow robots to provide human beings with a variety of services in uncertain and unstructured everyday environments [11].

In this section, we describe the UF service proposed for the integration and control of physical and virtual objects in the ubiquitous computing framework and introduce a robot control method supported by this service. The proposed UF service is composed of smart object, smart logic, and smart discovery services as shown in Fig. 2, and each service is implemented by using web services technology [5, 7].

2.1. UF Smart Object Service. Every object existing in the physical space has inherent functions. For example, a bulb can emit light, and a digital camera can capture an image of a physical object. Moreover, each physical object can be integrated with another to obtain a new function. For example, if we integrate a door, which is a simple object, an actuator, which is a dynamic object, and an infrared sensor, which can gather information, we can fabricate an automatic door to which we can add a speaker, which is an informative object, for improving safety and a face recognition system for improving security.

The UF smart object service defines such functions of physical objects as UFs in the virtual space of the ubiquitous computing environment and allows programmable resources on the computer networks to control the physical objects by using the UFs in virtual space.

UF can be defined simply as a characterized function of a physical object or as a composite UF, which is a combination of a few UFs. Thus, it is necessary to consider the interoperability of UFs. In order to address this problem, the proposed service classifies

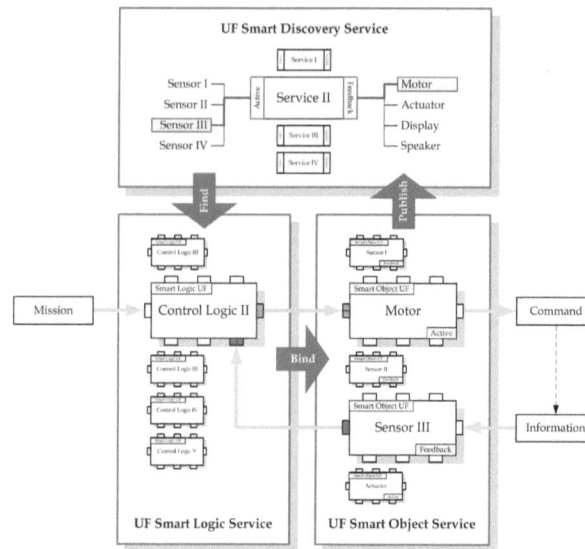

FIGURE 2. UF service.

physical objects on the basis of the signal change between the input and output channels of UFs. For example, an actuator/motor, which generates physical motion, and an audio/video device, which transmits information, are assigned an active UF that receives control signals through the input channel and transmits motion and information through the output channel. On the other hand, a sensor/camera is assigned a feedback UF that receives motion and information through the input channel and transmits the control signal through the output channel. Furthermore, a robot is assigned a composite UF that combines the abovementioned types of UF. However, a common object such as a book cannot posses an input and output channel by itself. Thus, common objects are associated with knowledge information in the virtual space by using a physical hyperlink such as an RFID tag, and information pertaining to these objects is transmitted through information UFs.

2.2. UF Smart Logic Service. An automatic door is physically composed of three components: the door, actuator, and infrared sensor. In order to define the relationship between these components from the viewpoint of distributed control and in order to control the door so that it opens when a robot approaches it, logic that controls the actuator through the feedback of information from the infrared sensor is required. Furthermore, in order for a robot to move from its current position to another position while performing a task, the robot requires a path generation algorithm to avoid collision with obstacles and a navigation algorithm to move to the target position along an the obtained path. Thus, the control algorithm of automatic doors and the path generation and navigation algorithms for mobile robots are components of the logic in the virtual space, and the services associated with such control logic are termed UF smart logic services.

The smart logic of the UF smart logic service integrates and controls the UF components of the service so that robots can complete their tasks. Fig. 3 illustrates the smart logic for automatic doors, where smart object UFs are associated with smart logic UFs on the basis of their input/output characteristics. Each UF block can be used in another design by linking it to other smart object UFs and smart logic UFs. Thus, the typical design of

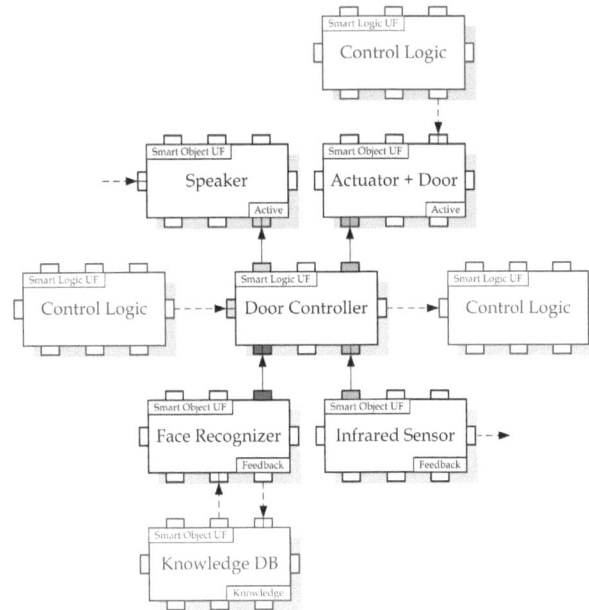

FIGURE 3. Smart logic UF and smart object UF for an automatic door.

robots using sensors, motors, links, and controllers can be extended through the recon-figurable combination of smart object UFs in the real domain and smart logic UFs in the virtual domain. This is an important advantage of ubiquitous robotics and might lead to the realization of the proposed service in ubiquitous computing environments.

2.3. UF Smart Discovery Service. The UF smart discovery service enables the discovery and integration of UFs so that programmable resources on the networks can utilize the UFs of physical objects in the everyday domain. In order to realize this service with the service oriented architecture (SOA) when the physical objects move, a wireless communication device that can send and receive information about the UFs and that has an easy interface for identifying mobile objects is required. For this purpose, we have developed the ubiquitous function activation module (UFAM), shown in Fig. 4.

The advantages of using this device include extension of the communication distance of the sensor network and low power consumption by the RFID. Furthermore, this device provides an IO port for digital input and output and an 8-bit microcomputer, which can control the sensor and actuator directly. If the microcomputer and the communication device are set in the standby mode using software and activated once every 5 seconds, the system can be operated on a single button battery for a year or longer. The size of the device is approximately $6.0\,cm^3$, making the device the smallest among battery-powered terminal in the world.

2.4. UF Service with SOA. From Fig. 2, it can be seen that the UF service has an SOA. A physical object integrated with the UFAM registers link information about its UF with the UF smart discovery service and the discovery service, in turn, provides this link information to the smart logic service when a robot utilizes UF services to perform a task. Subsequently, the smart logic service utilize the UFs to compose a control logic based on the characteristics of the UFs. If necessary, the UF smart discovery service provides this

FIGURE 4. UF activation module.

link information to client applications implemented on portable terminals such as PDAs and tablet PCs. Then the client applications compose a graphic user interface of the objects in the unknown space.

3. Smart Logic

In order to control a robot so that it performs its task successfully in the everyday domain, which is composed of various types of objects, smart logic should be used to assign UFs to the distributed objects accurately for analyzing the given task. However, changes in the characteristics of known objects and the appearance of unknown objects with time render this difficult and disturb the control system equipped with smart logic. Thus, robots require logic that can detect changes in objects in environments that vary with time and compensate for the disturbance caused by the changes. In this section, we discuss the smart logic based on a robust control algorithm, which provides flexible robustness to a robot functioning in the everyday domain.

3.1. Robust Internal-Loop Compensator. Two conditions must be satisfied in order to design a robust controller for a system with uncertainty and subjected to disturbance. The first condition pertains to robustness against uncertainties including external disturbance and dynamic changes in the system parameters, and the second condition to the performance specifications for the given task.

In the past few years, several studies have been performed to determine the manner in which these specifications can be met. The disturbance observer (DOB) [12], adaptive robust control [13], and enhanced internal model control [14] were developed in the course of these studies. The methods developed by these studies commonly require the design of a two-loop structure. This involves the design of an internal-loop compensator for improving robustness and the design of an external-loop controller for satisfying the given performance specifications. Fig. 5 shows the robust control system with a two-loop structure.

In such two-loop structure, the internal-loop compensator generates a corrective control input to reject the disturbance as much as possible so that an actual system can be adopted as the reference model. Here, the disturbance is defined as the sum of external disturbances and all possible disturbances due to differences between the actual and the

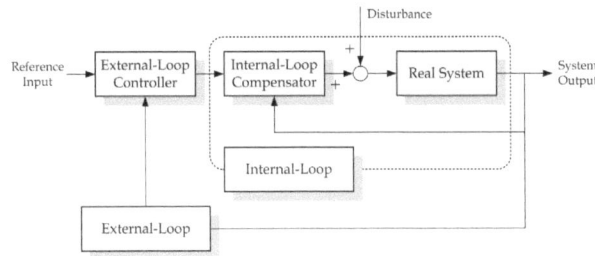

FIGURE 5. Robust control system with a two-loop structure.

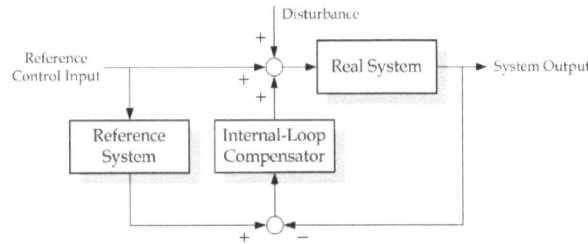

FIGURE 6. Robust internal-loop compensator.

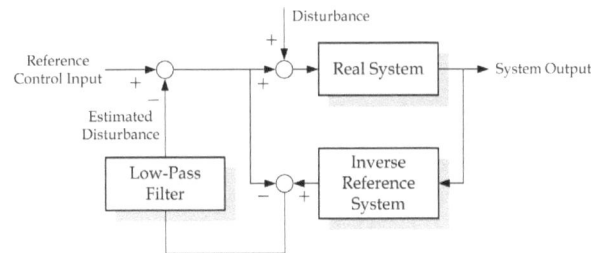

FIGURE 7. Disturbance observer.

reference system, such as modeling uncertainties and parameter variations. Hence, an actual system equipped with an internal-loop compensator can be regarded as the reference system if the internal-loop compensator works efficiently. On the other hand, the external-loop controller is designed to enhance the overall system performance, and the controller design is carried out by considering the reference system.

Recently, a generalized disturbance compensating framework named the robust internal-loop compensator (RIC), based on Lyapunov redesign, was developed. The RIC indicated the inherent structural equivalence of robust compensating controllers [15], and unified analysis and design were performed [16, 17]. Fig. 6 shows the structure of the RIC. In this figure, the difference between the reference system output and the measured real system output is defined as the model following error. Further, the control input is composed of the reference control input and the compensated control input.

Moveover, Fig. 6 can be equivalently transformed into the structure of a DOB. This indicates that if the internal-loop compensator is designed by considering the reference system in order to satisfy the given criterion, an optimal low-pass filter that has optimality

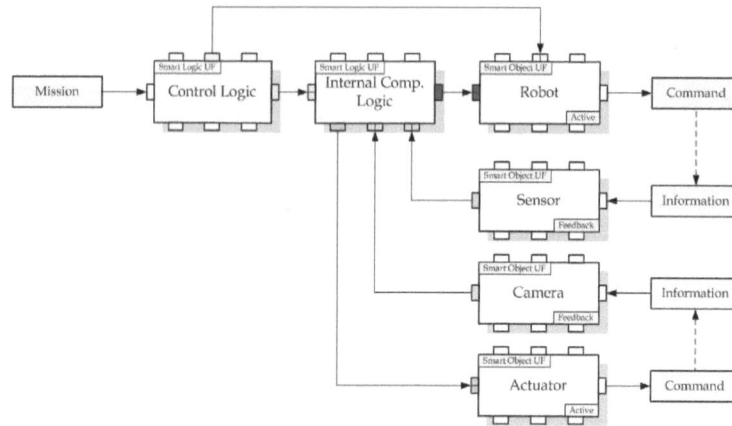

FIGURE 8. Smart logic with RIC structure.

under the given specifications is systematically designed, because the transfer function of the feedback system with the reference system and internal-loop compensator is the low-pass filter of the DOB. Therefore, the compensated control input of the RIC is identical to the estimated disturbance of the DOB [**18**]. This clearly indicates another important physical meaning immanent in the RIC structure.

3.2. Input and Output. In order to design smart logic with the proposed RIC structure, smart logic should be divided into control logic for improving performance and internal compensation (IC) logic for improving robustness and for accurately distinguish the disturbance of a given system, which explicitly defines explicitly the input and output of IC logic. This is because correlations between the tasks to be performed by the robot and the objects in the real world can be the source of the disturbance. Hence, disturbance is defined as the behavior of all objects that can be obstacle for robots in performing a task, and the feedback information obtained from the distributed objects is defined as the difference between the outputs of the reference and real systems in the RIC. On the basis of these definition, the input of the IC logic consists of the output of the feedback UF, which can detect disturbance, and the output of the control logic. Further, the output of the IC logic consists of the control inputs of the active UFs for disturbance compensation. As a result, the input and output of smart logic with the proposed RIC structure can be illustrated as shown in Fig. 8.

3.3. Smart Logic Based on the Reproduction of RIC Structure. The dynamic characteristics of a robot cause the tasks to be performed by it to be dynamic. Further, disturbance, as defined in the previous section, also has dynamic characteristics. Thus, the proposed structure shown in Fig. 8 should be generalized in order to compensate for the composite disturbance from dynamic environments that vary with time. For this purpose, we propose the reproduction of the RIC structure.

A control system with the RIC structure behaves like a reference system. Thus, the proposed RIC based method can be repeated to compensate for the dynamic characteristics of the composite system. The repetitive use of the RIC structure can eliminate disturbance in the whole system successfully and also minimize the changes in the performance caused by the dynamic characteristics. Fig. 9 shows the control method based on the proposed reproducible RIC structure.

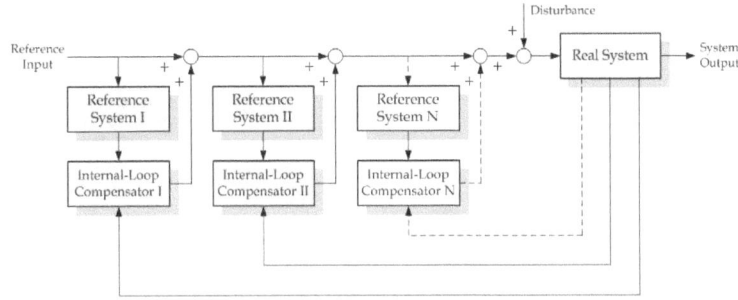

FIGURE 9. Reproducible RIC structure.

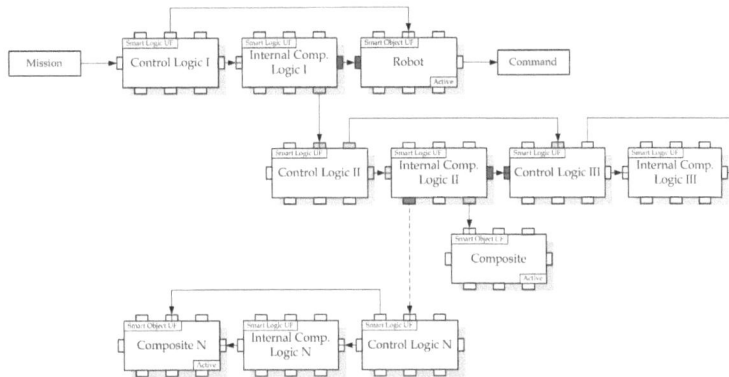

FIGURE 10. Smart logic with reproducible RIC structure.

The proposed method can be utilized for the smart logic design of a composite robot system with dynamic disturbance, as shown in Fig. 10. In this figure, the control logic for the given task uses its IC logic, which in turn uses another control logic to compensate for the disturbance, which acts as an obstacle in performing the task. Then, this control logic uses another control logic to perform the given task. Successive reproduction of the RIC structure enables the identification of the original obstacle and its elimination. The UF smart logic service utilizes this method by validating useful objects continuously through the UF smart discovery service with SOA. In the next section, we introduce a Ubiquitous Home using the proposed UF service and illustrate the control method for mobile manipulators based on the reproducible RIC structure.

4. Implementation

4.1. u-RT Space. We have implemented an RFID tag based system on the basis of the concept of universal design. Figure 11 shows the u-RT space for a robot with ambient intelligence.

In order to link the physical and virtual spaces, we utilized a physical hyperlink involving the use of two types of RFID tags. Figure 12 (a) shows the active type RFID tag, the UFAM, which is used to access Web services for obtaining the absolute position information of the space. Hence, a networked robot can download the map of the space using this RFID tag. Figure 12 (b) shows the passive type RFID tags that are implanted under the floor and identified using a tag reader attached to the bottom of the networked robot. This

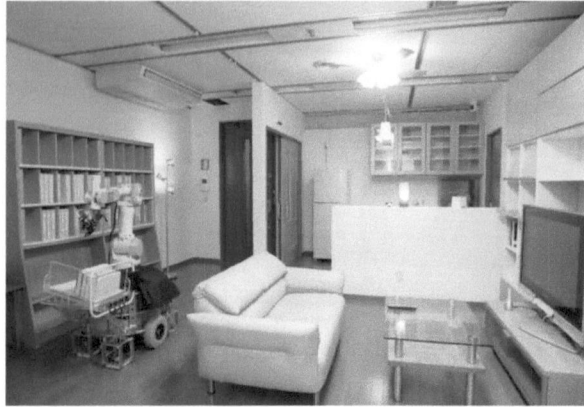

FIGURE 11. *u-RT space* with UFAM, floor tags, and mobile manipulators.

(a) (b)

FIGURE 12. Two kinds of RFID tags used for the u-RT space. (a) UFAM (active). (b) RFID tag and reader (passive).

RFID system is used for obtaining the relative position information of the u-RT space. Four hundreds RFID tags were implanted in the floor at 25 cm intervals along the x and y axes to allow reasonable localization, and their IDs and relative position data were stored in the virtual space as shown in Fig. 13. As a result, the networked robot can localize itself using position information obtained from the map of the u-RT space. Figure 14 shows the virtual space program used by the networked robot and u-RT space that supports the interface for accessing Web services. The virtual space program synchronizes the experiment system with the virtual space, which contains seven virtual cameras for remote monitoring.

For example, if the given task is to transfer the dishes on a table into a dishwasher, the sequence of ubiquitous localization and mapping based on RFID tags is as follows.

(1) The networked robot is given the task.
(2) The active tag of the networked robot communicates with the active tag of the u-RT space, and the passive tag reader obtains the current information from the floor tags.
(3) The active tags offer the ID, URL, and user information of the u-RT space, and the passive tags offer the ID, URL, and user information of the relative position.
(4) The networked robot retrieves the necessary knowledge from a data source from virtual Web resources.

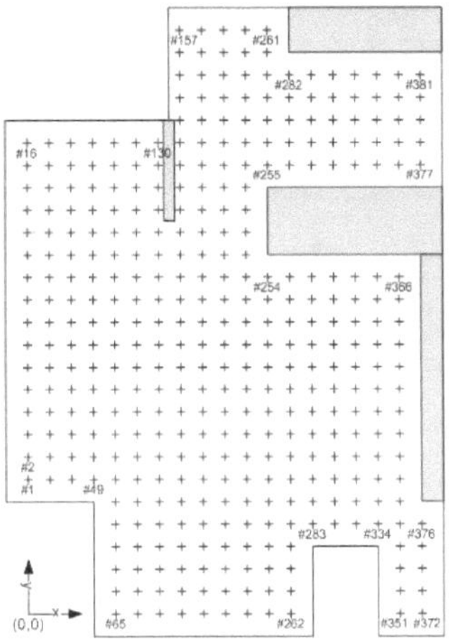

#	Tag ID	x	y
1	3843380000000C00	0	1.5
2	8A36380000000C00	0	1.75
3	4A23380000000C00	0	2
4	6D1C380000000C00	0	2.25
5	552A380000000C00	0	2.5
6	3327380000000C00	0	2.75
7	E532380000000C00	0	3
8	511B380000000C00	0	3.25
9	843B380000000C00	0	3.5
10	881C380000000C00	0	3.75
11	8031380000000C00	0	4
12	303A380000000C00	0	4.25
13	5823380000000C00	0	4.5
14	CC1E380000000C00	0	4.75
15	3E26380000000C00	0	5
16	6044380000000C00	0	5.25
17	8624380000000C00	0.25	1.5
18	7733380000000C00	0.25	1.75
19	4646380000000C00	0.25	2
20	0331380000000C00	0.25	2.25
...	...	...	...
...	...	...	...
...	...	...	...
375	CD3A380000000C00	4.5	0.75
376	8D1C380000000C00	4.5	1
377	0922380000000C00	4.5	5
378	3F23380000000C00	4.5	5.25
379	1242380000000C00	4.5	5.5
380	A538380000000C00	4.5	5.75
381	9623380000000C00	4.5	6

FIGURE 13. Map of the u-RT space and floor tag information.

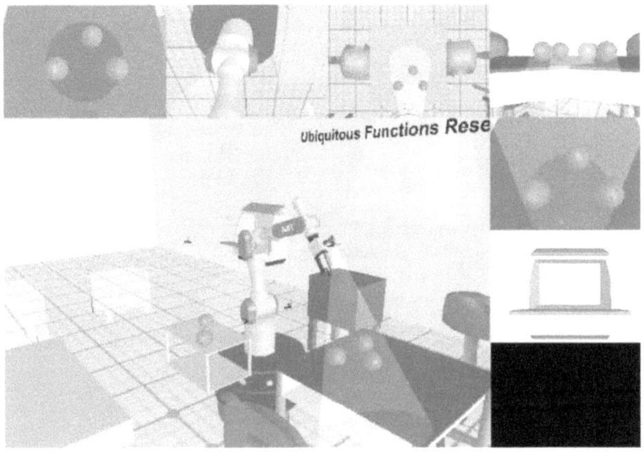

FIGURE 14. Virtual space program for networked robots and the u-RT spaces.

(5) The networked robot localizes the current position and composes the path by which to navigate to the target position.

(6) The networked robot navigates to the target position and compensates the path following error continuously by using the information from the space.

(7) If the networked robot arrives at the target position, if carries out the given task.

In order for a robot to perform a task in the u-RT space, the robot control logic should accurately assign the UFs of the distributed objects by analyzing the given task. However, changes in the characteristics of known objects and the appearance of unknown objects with time render this difficult and create disturbance. Hence, we applied the proposed UF services for the discovery and integration of objects and their components to resolve this problem.

Fig. 15 shows the control flow of the UF smart logic service for the abovementioned example. To enable the robot to carry out the given task, the dishwashing logic uses the IC UF. Then, the IC UF controls the robot so that it cleans the table by using the tableware UF and uses the path generation logic to enable the robot to move to the target position.

(a)

(b)

(c)

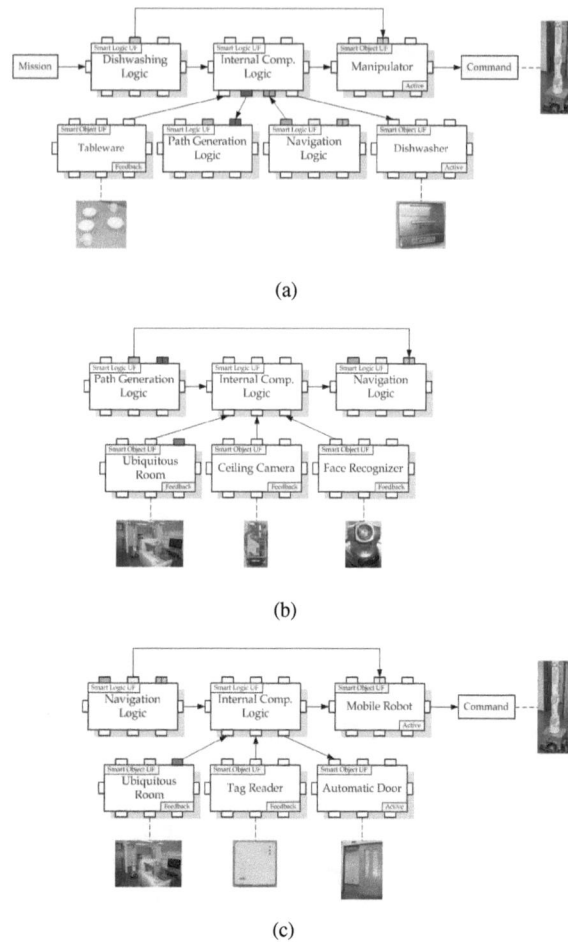

FIGURE 15. UF smart logic service for dishwashing. (a) Smart logic for Dishwashing. (b) Smart logic for mobile path generation. (c) Smart logic for mobile navigation.

Subsequently, the path generation logic uses the feedback UF of the current room and the objects that can be used for the task and transmits the navigation logic designed such that the robot path can compensate for the disturbance caused by tis generation. Then, the navigation logic controls the robot so that it moves to the target position by using the UF of the tag reader to read floor tags and an automatic door to open the door. If the robot arrives at the target position, the navigation logic transmits this information to the IC part of the dishwashing logic so that the robot arranges the dishes in the dishwasher by using the dishwasher UF and completes its task.

Fig. 15 (a) shows how the IC logic uses another control logic and receives a reply from the control logic in the context of the abovementioned example. Fig. 15 (b) shows how the IC logic is used when the control logic transmits its output to another control logic. Furthermore, Fig. 15 (c) shows how all the disturbance of the first IC logic is compensated through the successive reproduction of the RIC structure.

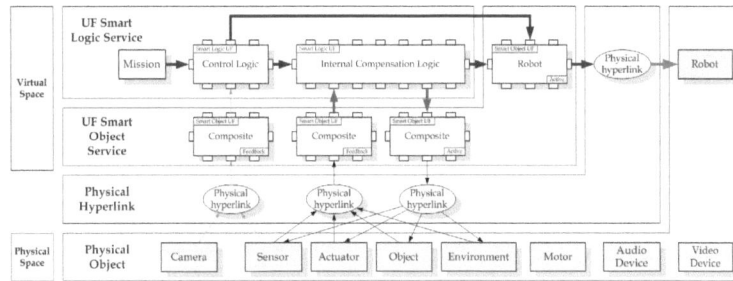

FIGURE 16. UF service for u-RT space.

FIGURE 17. Daily life space.

Finally, the UF service for the u-RT space is designed, as shown in Fig. 16. As can be seen in this figure, the mobile robot was controlled using only external sensors to examine the feasibility of the proposed method. In the case of the task mentioned above, the navigation error of the mobile robot was less than 5 cm, which is the detection range of the tag reader. This is a good result because the navigation logic UF uses only the floor tags and not the internal odometers of the mobile robot. The proposed method was tested on the librarian robot system of the *Intelligent Systems Research Institute of the National Institute of Advanced Industrial Science and Technology* [**19**]. It was found that the localization and mapping problem of robots can be feasibly solved using our method.

The current style of buildings is very functional. In particular, libraries and art museums can be easily structurized because their uses are clearly defined. The interiors of such buildings are well planned, and there are clear differences in the uses of each floor and each room. Further, because database systems for books and works of art have already been developed, these buildings are most suited for use as u-RT space. Hence, such buildings, which can be easily structurized, are the target spaces for the application of the proposed system.

4.2. Daily-life-supporting robots. Figure 17 shows a daily-living space, which was built so that daily-life-supporting robots carry out various tasks. In this application, a

(a)   (b)

FIGURE 18. Daily-life-supporting robots. (a) AIST robot. (b) Toshiba robot.

FIGURE 19. Ambient network camera systems.

wheelchair based robot with a manipulator, developed by AIST [20], and a service robot with two arms, developed by Toshiba, were used as daily-life-supporting robots, which are shown in Fig. 18.

In order to enable different kinds of robots to handle same objects in daily life environments, the following database table was designed based on the proposed universal design.

- ID
    Object RFID ID or CLUE ID
    Ex) 123ABCDEFGXXXX
- Name
    Object name
    Ex) Book (Lunch Box, Bottle, etc.)
- Category
    Category for affordance
    Ex) Portable (Appliance, Furniture, Space, etc.)
- RefCoord
    Type of reference coordinate
    Ex) Relative (Absolute)
- ZoneID
    Zone which an object belongs to

Ex) Mobile rack (Table, Robot, Bookshelf)

- Status

  Object physical status

  Ex) Open (Close, Occupied, Empty)

- ManipulationPosition

  Object position in zone local coordinate

  Ex) 150 0 0 0 0 1.57

  　　(x y z [mm] α β γ [radian])

- RotationPosition

  Rotation origin position in object local coordinate

  Ex) 0 0 0 0 0 0

  　　(x y z [mm] α β γ [radian])

- RotatableAngle

  Rotatable angle

  Ex) 6.28 6.28 6.28

  　　(α β γ [radian])

- RotatableRadius

  Radius of rotation or distance of movement

  Ex) 300 ([mm])

- HandType

  Robot hand type

  Ex) MJ/PJ (MJ: Multi-joint, PJ: Parallel-joint)

- GripPoition

  Grip position in object local coordinate

  Ex) 0 35 40 0 0 1.57 10

  　　(x y z [mm] a b g [rad.] depth [mm])

- CurrentPosition

  Object position in global coordinate

  Ex) 1000 20000 100 0 0 3.14

  　　(x y z [mm] α β γ [radian])

- Size

  Width, length, height from object local coordinate origin

  (min_w, max_w,min_l,max_l,min_h,max_h)

  Ex) -100 100 -150 150 -10 20 ([mm])

- Weight

  Object weight

  Ex) 600 ([g]) γ [radian])

- Material

  Object material

  Ex) Paper (Plastic, Ceramic, Wood, Steel, etc.)

- PicutreName

  Picture name

  Ex) Robotics.jpg

- PictureType

  Picture type

  Ex) JPEG (GIF, PNG, etc.)

- PictureLength

  Picture file length

(a)   (b)

FIGURE 20. Target objects based on universal design. (a) Universal handle. (b) Universal container.

          Ex) 17,353 ([byte])
- Picture
     Binary data for object image
     Ex) <Binary data>
- DateTime
     Date and time when data was updated lastly
     Ex) 2009-04-12 18:38:43.237

For equitable use of objects by robots, each item of the database table was determined. For flexibility in use, the hand type and the grip position item have multiple data sets. For simple and intuitive, the ID is represented as RFID/CLUE ID and the category is represented for basic affordance of an object [21]. For perceptible information, the reference coordinate, the zone ID, and the status of objects are added. The shape of daily objects was also designed by universal design so that different types of robots can manipulate it easily.

In order for the robots to realize a particular scenario, by which robots pick out a container in a freezer and move it to a table, the UF service was built based on the proposed design method. For localization of target objects, network cameras were installed as shown in Fig. 19, and the position data of objects were transmitted to the UF service. Figure 20 shows a universal handle and a universal container with CLUE(Coded Landmark for Ubiquitous Environment) [22, 23], respectively. If a robot recognizes an object CLUE, then the robot requests information for object manipulation and the UF service provides this to the robot. In order to verify robot behavior history, a visualizer of Fig. 21 was developed, which shows 3D graphic motions of a robot according to time using information that the UF service manages. In order to retrieve and update database information, the robots and the network camera systems call ubiquitous functions of the UF service, and the call process can be verified as shown in a Web application of Fig. 22. This web application shows the called ubiquitous functions, request messages, and return messages, which was realized based on AJAX (Asynchronous JavaScript and XML) technology.

5. Conclusion

In this chapter, seven principles of universal design with robots have been proposed and the UF service, composed of smart object, smart logic, and smart discovery services,

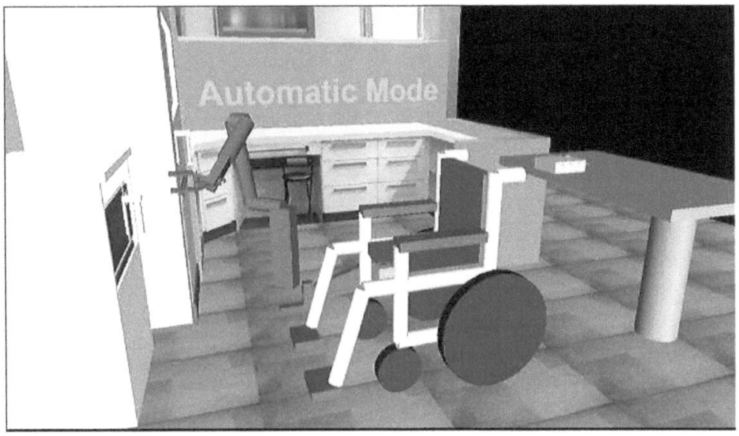

FIGURE 21. UF service 3D visualizer.

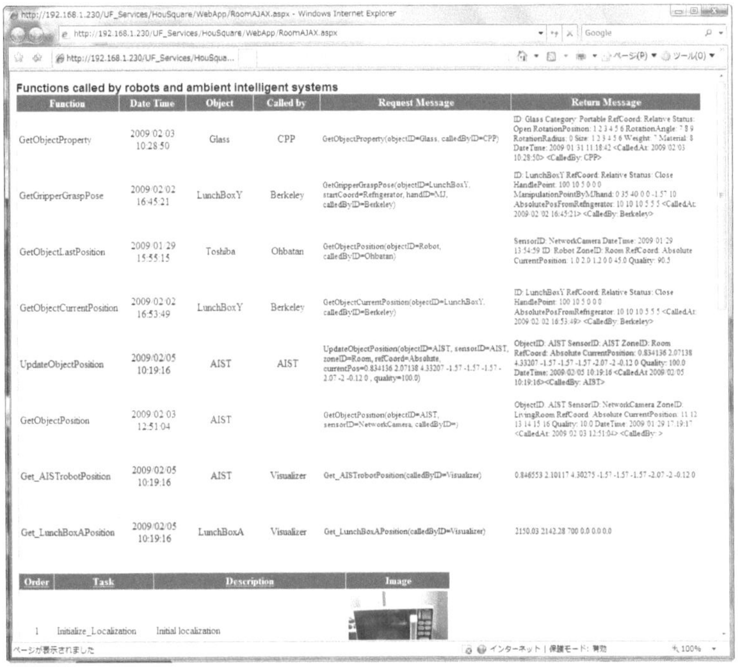

FIGURE 22. Web application which shows ubiquitous functions called by the robots and network camera system.

has been presented as a powerful method for the discovery and integration of objects and their components in the SOA. In addition, a reconfigurable control method for networked robots that can be realized in ubiquitous computing environments was discussed. To provide flexible robustness to the robot operating in a dynamic environment, the structural design of smart logic has been illustrated through using the reproduction of the RIC structure. The object database table was also designed such that different types of robots can handle target objects exactly in daily life environments.

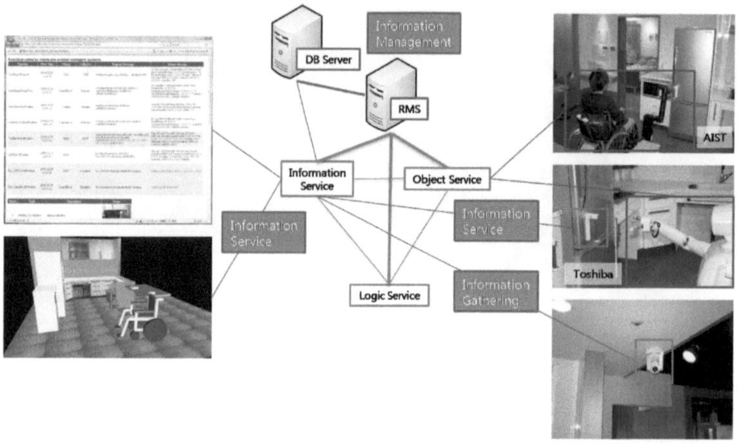

FIGURE 23. Overall control architecture realized by the UF service for daily-life-support robots.

Bibliography

[1] S. Helal, W. Mann, H. El-Zabadani, J. King, Y. Kaddoura, and E. Jansen, "The gator tech smart house: A programmable pervasive space." Computer **38**:50-60 (2005).

[2] A. Brooks, T. Kaupp, A. Makarenko, S. Williams, and A. OreAback, "Towards component-based robotics." Proc. 2005 IEEE/RSJ Int. Conf. Intelligent Robots and Systems (IROS2005) 3567-3572 (2005).

[3] F. Pont and R. Siegwart, "A real-time software framework for indoor navigation." Proc. 2005 IEEE/RSJ Int. Conf. Intelligent Robots and Systems (IROS2005) 3573-3578 (2005).

[4] N. Ando, T. Suehiro, K. Kitagaki, T. Kotoku, and W.-K. Yoon, "RT-middleware: Distributed component middleware for RT (robot technology)." Proc. 2005 IEEE/RSJ Int. Conf. Intelligent Robots and Systems (IROS2005) 3555-3560 (2005).

[5] B. K. Kim, M. Miyazaki, K. Ohba, S. Hirai, and K. Tanie, "Web services based robot control platform for ubiquitous functions.," Proc. 2005 IEEE Int. Conf. Robotics and Automation (ICRA2005) 703-708 (2005).

[6] B. K. Kim, K. Ohara, K. Ohba, T. Tanikawa, S. Hirai, and K. Tanie, "Networked robots in the informative spaces." Proc. Int. Conf. Control, Automation and Systems (ICCAS2005) 714-719 (2005).

[7] B. K. Kim, N. Tomokuni, K. Ohara, K. Ohba, T. Tanikawa, and S. Hirai, "Ubiquitous function services based control for robots with ambient intelligence." Proc. IEEE Industrial Electronics (IECON2006) 4546-4551 (2006).

[8] K. Ohara, K. Ohba, B. K. Kim, T. Tanikawa, S. Hirai, and K. Tanie, "Networked robots in the informative spaces." Proc. 2005 Int. Workshop on Networked Sensing Systems (INSS2005) 97-102 (2005).

[9] The Center for Universal Design, http://www.design.ncsu.edu/cud/index.htm.

[10] B. K. Kim, K. Ohara, K. Kitagaki, and K. Ohba, "Design and control of librarian robot system in information structured environments." Journal of Robotics and Mechatronics **21**: 507-514 (2009).

[11] N. Y. Chong, H. Hongu, M. Miyazaki, K. Takemura, K. Ohara, K. Ohba, S. Hirai, and K. Tanie, "Robots on self-organizing knowledge networks." Proc. 2004 IEEE Int. Conf. Robot. Automat. 3494-3499 (2004).

[12] H. S. Lee and M. Tomizuka, "Robust motion controller design for high-accuracy positioning systems." IEEE Trans. Ind. Electron. **43**:48-55 (1996).

[13] B. Yao, M. Al-Majed, and M. Tomizuka, "High-performance robust motion control of machine tools: An adaptive robust control approach and comparative experiments." IEEE/ASME Trans. Mechatron. **27**:63-76 (1997).

[14] H. A. Zhu, G. S. Hong, C. L. Teo, and A. N. Poo, "Internal model control with enhanced robustness." Int. J. Systems Sci. **26**:277-293 (1995).

[15] B. K. Kim, W. K. Chung, H. T. Choi, I. H. Suh, and Y. H. Chang, "Robust optimal internal loop compensator design for motion control of precision linear motor." Proc. 1999 IEEE Int. Symposium on Industrial Electronics 1045-1050 (1999).

[16] B. K. Kim, H.-T. Choi, W. K. Chung, and I. H. Suh, "Analysis and design of robust motion controllers in the unified framework." ASME J. of Dyn. Syst., Meas. and Contr. **124**:313-321 (2002).

[17] B. K. Kim and W. K. Chung, "Performance tuning of robust motion controllers for high-accuracy positioning systems." IEEE/ASME Trans. Mechatron. **7**:500-514 (2002).

[18] B. K. Kim and W. K. Chung, "Advanced disturbance observer design for mechanical positioning systems." IEEE Trans. Ind. Electron. **50**:1207-1216 (2003).

[19] "Introduction of ubiquitous robotics into home living environment : Build a living space with distributed robotic element devices utilizing wireless network and middleware techniques." http://www.aist.go.jp/aist_e/latest research_latest research.html.

[20] T. Tomizawa, J. H. Lee, Y.-S. Kim, Y. Sumi, H. M. Do, B. K. Kim, T. Tanikawa, H. Onda, K. Ohba, and T. Sonoyama, "Design of common software architecture for multi robot platform - the concept and the system architecture of an electric wheelchair -." Proc. 2008 SICE System Integration (SI2008) 469-470 (2008).

[21] S. Hidayat, B. K. Kim, and K. Ohba, "Learning affordance for robots using ontology approach." Proc. 2008 IEEE/RSJ Int. Conf. Intelligent Robots and Systems (IROS2008) 2630-2636 (2008).

[22] K. Ohara, T. Sugawara, J. H. Lee, T. Tomizawa, H. M. Do, X. Liang, Y. S. Kim, B. K. Kim, Y. Sumi, T. Tanikawa, H. Onda, and K. Ohba, "Visual mark for robot manipulation and its RT-middleware component." Advanced Robotics **22:**633-655 (2008).

[23] T. Tomizawa, J. H. Lee, Y.-S. Kim, H. M. Do, Y. Sumi, B. K. Kim, T. Tanikawa, H. Onda, K. Ohba, and T. Sonoyama, "Common interface design for robot and human -universal handle for object manipulation-." Proc. 2008 SICE System Integration (SI2008) 465-466 (2008).

CHAPTER 9

Ubiquitous Robotic Space and Its Real-world Applications

Wonpil Yu, Jae-Yeong Lee, Heesung Chae, Yu-Cheol Lee,
Minsu Jang, and Joo-Chan Sohn

Robot Research Department

Electronics and Telecommunications Research Institute

Daejeon, Korea

Hyosung Ahn

Department of Mechatronics

Gwangju Institute of Science and Technology

Gwangju, Korea

Young-Guk Ha

Department of Computer Science and Engineering

Konkuk University

Seoul, Korea

and

Yong-Moo Kwon

Imaging Media Research Center

Korea Institute of Science and Technology

Seoul, Korea

ABSTRACT. This paper is an extension of the authors' earlier publication proposing *ubiquitous robotic space* [1]. In the previous work, authors collaborated to propose a novel robotic service framework. The proposed robotic service framework comprises three conceptual spaces: physical, semantic, and virtual space, which we call a ubiquitous robotic space collectively. We implemented a prototype robotic security application in an office environment, which confirmed that the proposed framework is an efficient tool for developing a robotic service employing IT infrastructure, particularly for integrating heterogeneous technologies involved and dealing with inherent complexity found in robotic platforms. The proposed ubiquitous robotic space architecture is further applied to a large scale artificial environment, which is crowded with hundreds of visitors. This experimental trial proves further the usefulness of the proposed architecture since virtually no modification of the original service architecture is required.

1. Introduction

Until recently, robots have demonstrated their usefulness mainly for industrial applications [2]. These applications typically carry out a series of predetermined operations in a precisely controlled environment such as a manufacturing site. Today, there is a growing interest for using robots beyond the manufacturing site to the everyday environment and to accommodate various demands raised as social, industrial, and personal needs change. The above situation makes robot tasks formidable since structuring the environment is subject to restriction and unsupervised understanding thereof is hard to achieve.

While traditional robotics research focused on enhancing functionalities of robots, a new computing paradigm, which is now called *ubiquitous computing*, has been proposed [3]. To date, a number of novel technologies have been proposed supporting the idea of ubiquitous computing: radio frequency identification, wireless sensor network, mobile devices, broadband convergence network, etc. Small networked computers with a sensing functionality lies at the heart of ubiquitous computing, embedded in everyday objects to serve people invisibly and unobtrusively in the background [4].

Ambient intelligence is a different representation of ubiquitous computing paradigm, emphasizing distributed sensing embodied in our daily environments. Ambient intelligence entails structuring the environment into a smart space which can recognize and respond to the needs of individuals. EasyLiving, Intelligent Classroom, MavHome [5], and iRoom projects are some of recent developments for smart spaces in the context of ambient intelligence [6].

Likewise, in the robotics field, many researchers have paid attention to building smart spaces. While the primary concern of the smart space in the original context of ambient intelligence is to support humans, those incorporating robotics attempt to enhance recognition and interaction capabilities of a robot as well as supporting humans, thus extending usefulness of the robot in the daily environment.

In that regard, intelligent space has been proposed where the space supports both humans and robots by implementing Distributed Intelligent Network Device (DIND) architecture [7, 8]. Robotic Room [9, 10] describes research works to design and implement distributed robotic components or devices to assist humans living inside the environment. Omniscient space [11] is a research work to connect heterogeneous knowledge resources to build up task intelligence of a robot. More recently, WABOT-House project has been initiated to construct buildings providing a living space for humans and robots [12].

Compared with the developments of ambient intelligence, those robotics-oriented spaces differ in that they necessarily entail precise location information of robots and objects including humans, as evidenced in the traditional robotics community [**13**]. In addition to various aspects to consider when deploying a location sensing infrastructure [**14–17**], robot-specific features such as mobility and manipulability should be taken into account when designing and deploying a sensor network. References [**18–20**] report synergistic uses of sensor networks and mobility of robots, illustrating how traditional robot tasks can be reformulated based on distributed sensing strategy.

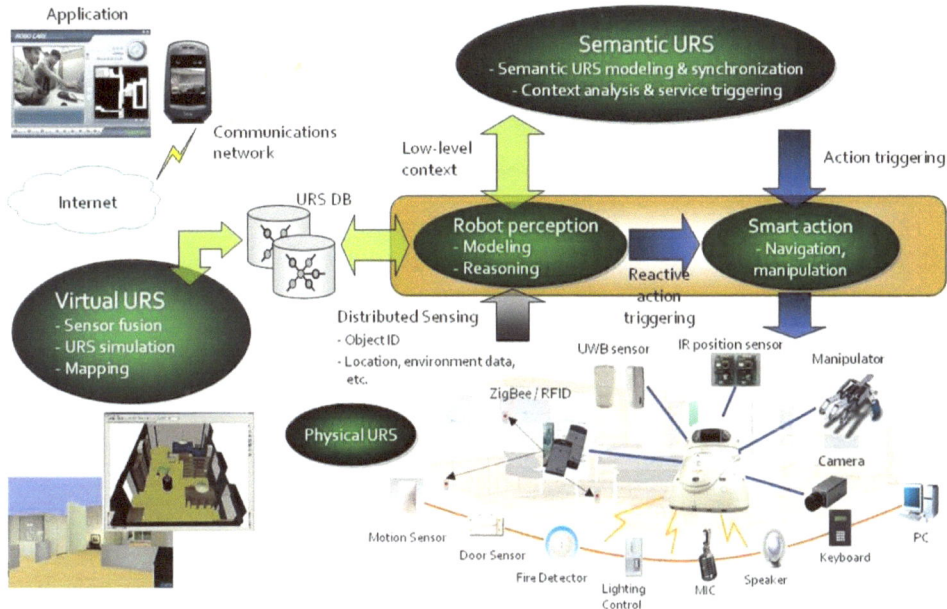

FIGURE 1. Conceptual structure of a ubiquitous robotic space.

Since the implementation of the smart environment requires integration of a large number of heterogeneous components, architectural issues for designing and building a smart environment have also received much attention. In this regard, a prototypical configuration of a networked robot system is described in [**21**]. However, reference [**21**] paid more attention to the control of networked robots rather than architectural issues of building a smart environment.

The aforementioned DIND is one of early works for converting an everyday environment into an intelligent space. It was suggested that DIND meets the requirements of intelligent space characterized by scalability, reconfigurability, modularity, and easy maintenance [**7, 8**]. However, DIND is more concerned with a sensor network missing various aspects of intelligent space: task control, user interface, world modeling, etc.

Reference [**22**] describes an approach for utilizing networked resources available to build a robotic system. Authors of [**22**] have introduced the concept of a *module* which represents a networked resource and acts as an interface between other modules. They also called a collection of modules *module pool*, to which a task module broadcasts queries specifying its given task; dynamic reconfiguration utilizing networked resources and task description is a main feature characterizing the proposed work.

Physically Embedded Intelligent System (PEIS) ecology described in [**23–25**] aims to build a smart environment characterized by ambient intelligence and networked robot systems. The four basic functionalities used to build a PEIS element are M (Modeling), D (Deliberation), P (Perception), and C (Control), the structure of which follows the hybrid paradigm of robot control architecture [**26**].

Along with the introduction of PEIS-ecology, authors thereof have pointed out several challenging issues: dynamic regime, heterogeneity, and anchoring [**25**]. To deal with the respective issues, various methodologies have been proposed such as hybrid configuration of PEIS-ecology, distributed communication based on tuple space combined with an event mechanism [**23**], and cooperative anchoring framework to fuse different types of information provided from various PEIS components.

A recent work deals with an intelligent space incorporating networked robot systems from control point of view [**27**]. Three ubiquitous function (UF) services have been proposed comprising UF smart object service, UF smart logic service, and UF smart discovery service; the functional role of a task module introduced in [**22**] corresponds to that of UF smart logic service. The authors have employed web service technology to implement each UF service.

In this paper, we propose a ubiquitous robotic space (URS) which refers to a special kind of environment in which robots gain enhanced perception, recognition, decision, and execution capabilities through distributed sensing and computing, thus responding intelligently to the needs of humans and current context of the space.

The proposed URS also aims to build a smart environment by developing a generic framework in which a plurality of technologies including robotics, network and communications can be integrated synergistically. Fig. 1 illustrates the conceptual structure of the proposed ubiquitous robotic space.

The proposed URS comprises three spaces: physical, semantic, and virtual space. The current implementation of the physical space consists of a localization network for assisting robot navigation, a mobility-supporting wireless sensor network, and a URS server that manages various physical components including the two networks. The semantic space provides two functionalities of situation understanding and service generation in accordance with the interpretation of the situation. The virtual space enables the user to interact with the physical space.

As pointed out in [**14, 15**], implementation and maintenance of precise localization network in an actual environment is a formidable task. In fact, localization data may influence the overall performance of a robot system making use of mobility of a robot. Mobility of a robot should also be considered when a wireless sensor network is to be utilized; automated data routing according to the current position of a robot is not well-established yet. Visualization and control of the physical world is also a non-trivial issue related to creating a user-friendly interface to the intelligent environment. Mapping and simulation of the physical space and interaction therewith are categorized into the realm of the proposed virtual space.

The remainder of the paper is organized as follows. Section 2 describes basic elements of the physical space and characteristics of each individual element. Section 3 introduces the semantic space and its constituting elements. The virtual space is described in Section 4, where described are a procedure for modeling environments and an implemented user interaction means to control physical elements in the physical space. A robot security application making use of the proposed architecture of the ubiquitous robotic space is described in Section 5. Section 6 concludes the paper and introduces further works.

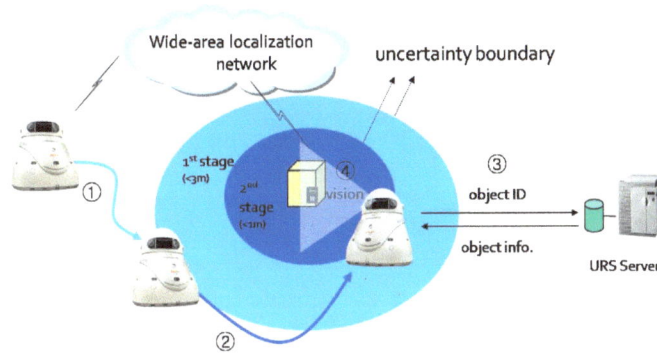

FIGURE 2. Two-stage localization scheme.

2. The Physical Space

Wireless sensor network and localization network are two fundamental components supporting the physical space, influencing the behavior of a robot inside the physical space. For managerial point of view, combining the two heterogeneous networks into a single physical network looks more economical. Restrictions on cost, node complexity, and operation mechanism, however, favors separation of the two networks.

Another component essential for management of the three spaces and data flow among them is a URS server. The URS server also handles various connectivity works between users and the Internet or existing communications network.

In what follows, the two networks that we have developed are described in detail.

2.1. Localization Sensor Network. The localization network tries to embody human behavior to localize an object. Fig. 2 illustrates the idea embodied in the proposed localization network. At first, a mobile robot obtains the location information of an object in question from a localization network providing wide coverage but low accuracy (*e.g.*, a ZigBee network). Localization is carried out based on RSSI and thus, the uncertainty of the location information may readily exceed two meters. The robot approaches the vicinity of the object and again localizes the object by utilizing a precision localization sensor. At this stage, we may optionally get detailed information about the object through RFID tag reading. The robot may then calculate posture of the object by using a vision sensor, after which manipulation of the object, for example, is carried out.

The aforementioned scenario resembles human behavior when we look for something with imprecise information at the initial stage. Reference [**28**] provides detailed description of the wireless localization network intended to be used at the initial stage. In this paper, we describe a precision localization sensor suite which is used to narrow down the uncertainty of the initial stage to meet the requirements of robot navigation.

Each localization sensor suite comprises two infrared beacon modules attached on the ceiling and an image sensor equipped on top of a mobile robot. This configuration is, in fact, very well recognized in robotics community [**29–31**]. As pointed out in [**14**], however, several constraints should be satisfied if any localization technique is to be a practical solution to real-world robotic applications, particularly for building an intelligent space incorporating robot navigation. The following are the constraints that we have considered to build a precision localization network:

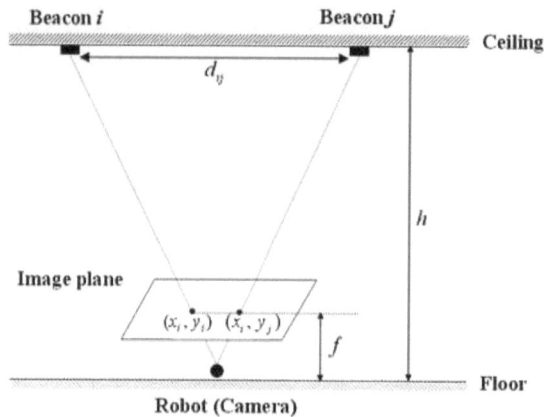

FIGURE 3. Sensor configuration for estimating robot location. f, d_{ij}, and h represent camera focal length, distance between two neighboring infrared LEDs, and height measured from the camera optical center to the ceiling, respectively.

- Accuracy: for the convenience of developing robot navigation tasks, the positional accuracy should be less than 20 cm in x, y each, orientation accuracy being less than 5°.
- Repeatability: for a mobile robot to navigate reliably, jitter in the location information should be bounded by 1 cm of positional error variance and 1° of orientation error variance.
- Coverage: for a localization network to be meaningful, its coverage should be unlimited, which means that the localization network must be highly scalable.
- Response time: update frequency for locating a robot should be high enough. It was set to be more than 10 Hz to provide 3-D localization information (x, y, θ).
- Availability: location information about a robot should be provided at any time of the day and any place of the indoor environment.
- Deployment: the localization network should be capable of wireless operation except for power supply.
- Cost: constituting elements should be cost effective for the proposed localization network to be deployed in a large indoor environment.

We have chosen an optical tracking scheme to realize a localization network satisfying the constraints above. Although a localization network can be realized by using other means [13, 15], optical triangulation was found most excellent for this purpose mainly due to fast response under line-of-sight condition, high accuracy, negligible jittering, and cost. Other constraints could be met by making efficient use of the localization system. Fig. 3 is a basic sensor configuration to build the localization network.

The proposed sensor suite is configured such that infrared beacon modules are attached on the ceiling and an image sensor is mounted on top of a mobile robot as shown in Fig. 3. The image sensor is oriented to look upward so that its optical axis is perpendicular to the ground. For the sake of maximal field of view, a wide-angle camera lens is utilized. Each beacon module contains an infrared LED where we can control the on-off status of the LED wirelessly by using a unique identifier defined for each beacon module.

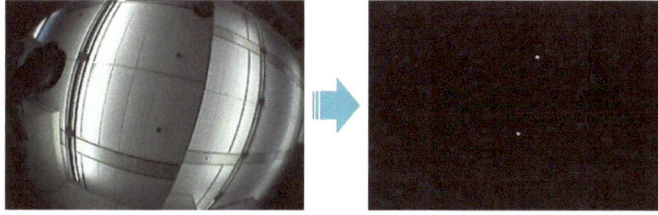

FIGURE 4. Effect of using a bandpass filter in the image capture step.

The localization is performed in three steps: image capture, blob extraction, and location calculation. In the image capture stage, a bandpass filter allowing only near-infrared light is employed to simplify computational complexity of the succeeding steps. Fig. 4 illustrates the effect of using a bandpass filter in the image capture stage.

In the second step, the captured image is dichotomized by using a threshold value determined in an adaptive manner. Connected component labeling is then applied to the dichotomized image to extract image blobs of LEDs. The position of each LED in the image is specified by the center of mass of each individual blob:

$$(1) \qquad x_k = \frac{1}{S_k} \sum_{(x,y) \in b_k} x I(x,y),$$

$$(2) \qquad y_k = \frac{1}{S_k} \sum_{(x,y) \in b_k} y I(x,y),$$

where (x,y) is a pixel coordinate; $I(x,y)$ is an intensity value at (x,y); b_k is k-th blob; and $S_k = \sum_{(x,y) \in b_k} I(x,y)$.

Finally, we can obtain the location data of a mobile robot by using some geometric calculations. For the purpose of simplicity, the image center is assumed to be aligned with the rotating axis of the robot; one can accommodate an offset of the image center from the robot's rotating axis by introducing a simple translation vector. The arrow originating from the image center in Fig. 5 represents the forward direction of the robot. L_i and L_j represent the position vector of two tags, the world coordinates of which are known beforehand. We can get image coordinate data of L_i and L_j from the previous blob extraction step.

Now, the localization is to calculate the translation vector from the origin of the world coordinate frame and orientation angle of the image coordinate frame measured from the world coordinate frame. The following are the equations needed for localization.

$$(3) \qquad s = \frac{D}{d} = \frac{\sqrt{(X_i - X_j)^2 + (Y_i - Y_j)^2}}{\sqrt{(x_i - x_j)^2 + (y_i - y_j)^2}},$$

$$(4) \qquad \cos\theta = \frac{(x_j - x_i)(X_j - X_i) + (y_j - y_i)(Y_j - Y_i)}{Dd},$$

$$(5) \qquad \sin\theta = \frac{(y_j - y_i)(Y_j - Y_i) - (x_j - x_i)(X_j - X_i)}{Dd},$$

$$(6) \qquad \begin{bmatrix} r_x \\ r_y \end{bmatrix} = s \begin{bmatrix} \cos\theta & -\sin\theta \\ \sin\theta & \cos\theta \end{bmatrix} \left(\begin{bmatrix} c_x \\ c_y \end{bmatrix} - \begin{bmatrix} x_i \\ y_i \end{bmatrix} \right) + \begin{bmatrix} X_i \\ Y_i \end{bmatrix},$$

where (X_i, Y_i) is the world coordinate and (c_x, c_y) is the coordinate of image center.

In Eq. 3, D and d denote distance values measured with respect to the world coordinate frame and image coordinate frame, respectively. Since we know the world coordinate data

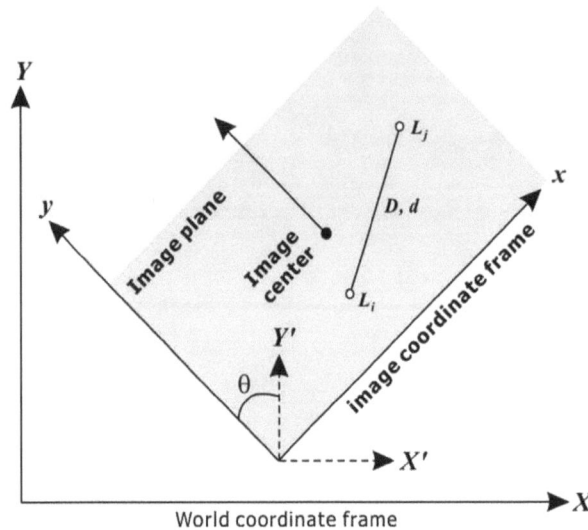

FIGURE 5. Definition of image and world coordinate frame.

of L_i and L_j; and image coordinate data of L_i and L_j from blob extraction step, we can calculate the scale variable s as shown in Eq. 3.

The achieved performance of the proposed localization sensor is that position and orientation error are less than ± 5 cm and $\pm 1°$. Repeatability error is confined to be less than 1 mm and $0.01°$. The achieved maximum update rate of location data is 30 Hz and coverage area is approximately a circle of radius 2 m with a ceiling height of 2 m.

Fig. 6 illustrates the developed precision localization sensor (which is called *Star-LITE*) and the corresponding schematic diagram showing the internal function blocks. We used a low cost fixed-point DSP for the detector where the above image processing and localization algorithm are implemented. Only serial interface is needed to install the detector in a robot.

2.2. Wireless Sensor Network. The sensor network platform to build a ubiquitous robotic space is called u-Clips and composed of sensor node hardware and sensor network protocol software. The u-Clips sensor node was designed to meet common requirements that a wireless sensor network platform should satisfy: small size and low power consumption. Also, it was so designed that various kinds of environmental sensors can be easily mounted to the sensor node hardware.

We have developed the u-Clips sensor network protocol stack based on Zigbee network protocol [**32**], which is a de facto standard wireless sensor network protocol due to its support for low power and reliable data communications. In addition to the standard Zigbee network protocol, the u-Clips sensor network protocol stack provides a novel mobility-supporting procedure to support mobility of a mobile robot equipped with a wireless sensor or sink node.

1) u-Clips Sensor Node. The developed sensor node is 3.5 cm × 5.5 cm in size and employs CC2430 Zigbee SoC from Chipcon. The CC2430 merges an 8051 MCU core with a 2.4 GHz RF transceiver into a single chip, providing excellent performance in terms of power consumption compared with the case of using separate components.

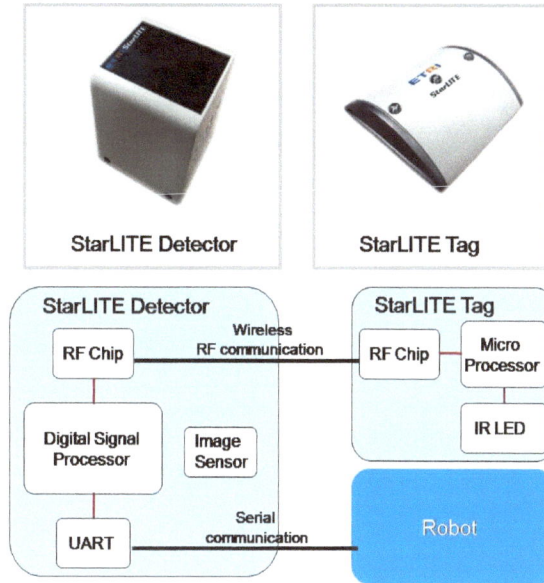

FIGURE 6. The image and schematic diagram of the developed precision localization sensor suite (detector and tag).

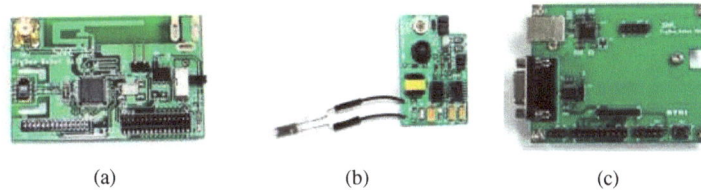

FIGURE 7. Hardware components of u-Clips sensor node. (a) Processor board. (b) Sensor board. (c) Programming board.

As shown in Fig. 7, the sensor node hardware consists of three components: a processor board, a sensor board and a programming board.

The processor board includes a CC2430 main processor, 2.4 GHz PCB antenna, and SMA antenna connector. It also accommodates a couple of basic sensors such as temperature, humidity, light sensor, and magnetic door sensor.

The sensor board is piggy-backed on the processor board and provides a dedicated sensing functionality, interfacing more specialized sensors such as PIR (Pyroelectric InfraRed) sensor, CO gas sensor, and flame sensor.

The programming board provides a USB and RS232C interface for rendering the processor board into a sink node. A processor board is piggy-backed onto the programming board to be connected to a robot and provides the robot with sensor data collected from each individual sensor node. In general, a sink node is connected to a fixed base station (for instance, the URS server). However, in order to better utilize the mobility of a robot, a u-Clips sink node is installed inside a mobile robot in our approach.

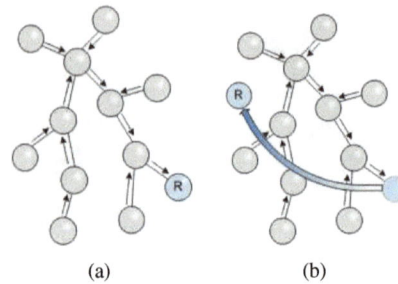

FIGURE 8. (a) Tree routing algorithm. (b) Mobility problem.

2) u-Clips Sensor Network Protocol. The base network protocol used in the u-Clips sensor network protocol stack is Zigbee. Zigbee is an emerging standard for wireless PAN (Personal Area Network) based on IEEE 802.15.4 standard [33]. Thanks to the addressing scheme of Zigbee, data from sensor nodes can reach the robot by using the tree topology without any routing table as illustrated in Fig. 8-(a). As can be noticed, tree routing requires little memory overhead and can also significantly reduce route discovery overhead compared with a table-driven mesh routing algorithm.

Different from existing sensor network applications, a mobile robot equipped with a wireless sensor or sink node introduces a peculiar feature to the conventional sensor network. As shown in Fig. 8-(b), if the robot moves out of RF transmission range of its current parent node, it cannot communicate with the entire sensor network. However, it is an essential requirement for a sensor network collaborating with robots to guarantee seamless communication with the mobile node by managing the mobility.

Currently, Zigbee specification version 1.0 [32] does not support node mobility. To support the mobility of a mobile robot, we propose a novel mobility-supporting procedure in which a node installed on a mobile robot (henceforth called a mobile node) should behave as an end device in a Zigbee network. In other words, a mobile node must not have any child nodes in a sensor network. It is because if the communication link between the mobile node and its parent node is broken as the robot moves, then all the data transmission to or via the mobile node will be disabled.

Out of transmission range, the mobile node will try to re-establish its communication link to a nearby node at the rejoining phase of the proposed procedure. If the mobile node rejoins the sensor network successfully, it gets a new network address according to the Zigbee specification. In this case, sensor nodes retaining a previous network address of the mobile node cannot communicate with the mobile node. So, we should also take into account how to announce a new network address of the mobile node to the entire sensor network.

Fig. 9 illustrates the proposed mobility supporting procedure of u-Clips sensor network using standard Zigbee protocol primitives. The proposed mobility supporting procedure consists of a link failure detection phase, network discovery phase, rejoin phase and announcement phase. The following will explain each phase in more detail.

Link failure detection phase: The mobile node (child node) periodically sends 'hello' message to its parent node to detect link failure. If acknowledgements are missing more than MaxRetries times, the NWK (network) layer of the mobile node notifies its APL (application) layer that the link to the parent node has been broken by calling NLDE-DATA.confirm primitive with a parameter value NO ACK.

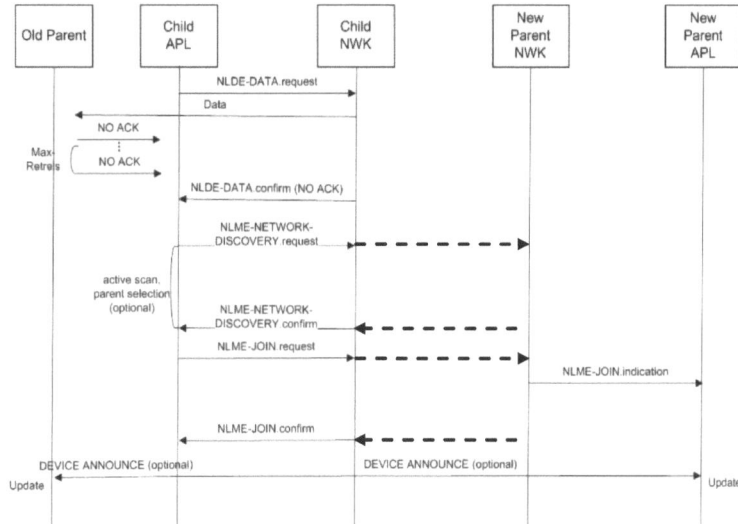

FIGURE 9. Proposed mobility supporting procedure.

Network discovery phase: The application layer tries to discover potential parent nodes around the mobile node by calling NLME-NETWORK-DISCOVERY .request primitive. During the discovery phase, the application layer can find a list of potential parent nodes and their PAN IDs, tree levels, LQIs (Link Quality Indications), and so on. Then, the application layer chooses a potential parent node that has the smallest tree level and a sufficient LQI value among nodes that have the same PAN ID. The reason why the mobile node selects the node of the smallest tree level is to reduce the routing cost; by doing so, the number of routing hops will be reduced because the network topology follows a hierarchical tree.

Rejoin phase: After choosing a new parent node, the mobile node requests to rejoin the network by calling NLME-JOIN.request primitive. Then the network layer of the mobile node transmits an ASSOCIATION_REQUEST command to the new parent node. The parent node then accepts the request for association if the number of current child nodes is less than a predefined maximum number. If the parent node accepts the joining request of the mobile node, it will assign a new 16 bit network address to the mobile node and send back an ASSOCIATION_RESPONSE command along with the newly assigned network address.

Announcement phase: After the mobile node successfully rejoins the new parent node, it broadcasts its new 16 bit network address to the entire sensor network using a DEVICE_ANNOUNCE command. Since the newly assigned network address of the mobile node is notified to all the sensor nodes in the network, sensor data can be seamlessly routed to the mobile robot. If the mobile node fails to rejoin the sensor network, it repeats discovery and rejoin phases until it succeeds to associate with a new parent node.

3. The Semantic Space

From the perspective of the traditional SENSE-PLAN-ACT control architecture, the semantic space resembles the planning phase [26]; it collects the sensory data from the

Context-Aware Service Execution (CASE)

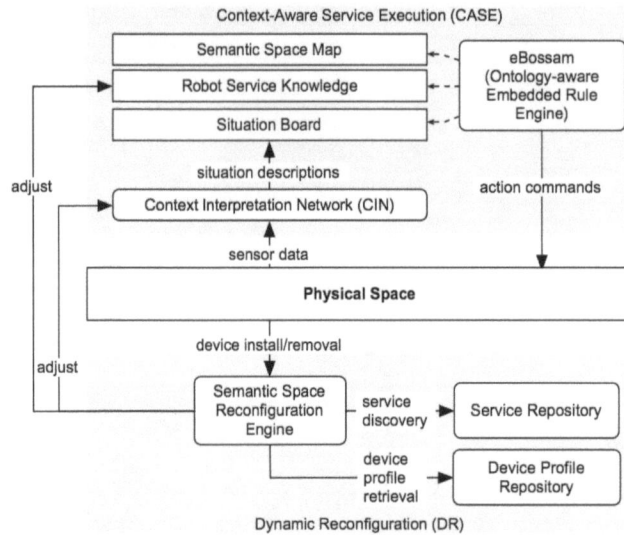

FIGURE 10. The semantic space consists of two main function modules: context-aware service execution (CASE) module and dynamic reconfiguration (DR) module.

physical space, builds up a world model, decides actions to trigger, and issues control commands back to the physical space.

Another important feature of the semantic space is related to heterogeneity and dynamic nature of the physical space; the semantic space collects device profiles of the sensors in the physical space, discovers the functionalities of the sensors, and deploys the corresponding service components for interpreting and utilizing the sensory data. A mobile robot, in this case, can be regarded as a mobile node providing multi-modal sensory data.

To summarize, two main functions of the semantic space are context-aware service execution and dynamic reconfiguration. Fig. 10 shows the overall structure of the semantic space and how it is interfaced with the physical space.

3.1. Context Aware Service Execution (CASE). The role of the CASE module is to generate action commands based on the situation described by the world model which consists of three symbolic data models: semantic space map, situation descriptions stored in the situation board, and robot service knowledge.

The semantic space map describes the spatial configuration of the physical space. It is a knowledge base containing symbols and formal sentences, each of which describes a specific physical position or region. The following sample is a set of sentences about a physical region, which states that the region bounded by (0, 45, 30, 75) is a meeting room and it's a restricted area.

```
region001 rdf:type   s:MeetingRoom;
          rdf:type   s:RestrictedArea;
          s:x        "0"^^xsd:integer;
          s:y        "45"^^xsd:integer;
          s:width    "30"^^xsd:integer;
```

```
s:height    "30"^^xsd:integer.
```

The situation description is a set of formal sentences that describe current status of the physical space, e.g., the position of a robot, atmospheric status of the corresponding region, entrance and exit events, etc. These sentences are dynamically generated by a context interpretation network (CIN).

The CIN consists of a set of hierarchically layered context interpreters. Figure 11 is a simple example of a CIN that generates descriptions of atmospheric mood from sensory data about temperature and humidity.

Each circular node, called a context interpretation node, consists a set of context interpretation rules that are used to perform context conceptualization or synthesis. The rectangular nodes represent input, output, or both contexts.

In our case, the CIN is dynamically constructed by a dynamic reconfiguration module described in Section 3.2 at the time of installation or removal of sensor devices comprising the u- Clips sensor network. The CIN continuously generates a series of output contexts which are stored in the situation board.

The situation board is a list of (k, v) pairs where k is context descriptor and v is the value of the corresponding descriptor. For example, the temperature value of 26°C at a location identified by a symbol LOC1 is stored as a pair, (<Temperature, LOC1>, 26). If the temperature of LOC1 changes, the value of the pair is updated to reflect the change. As such, the content of the situation board is a snapshot of changing situation of the physical space.

The robot service knowledge includes a set of service rules that trigger action commands based on the situation descriptions. The service rules also make use of the semantic space map for encoding spatial properties of a service. The following is a sample service rule that detects intrusion in a restricted area.

```
rule intrusion-detection is
if
  EntranceLocation(?ent,?loc)
  and RestrictedArea(?loc)
  and (not actor(?ent,?act) or
       (actor(?ent,?act)
        and not admitted(?loc,?act))
        )
then
  Intrusion(?ent,?loc);
```

The rule generates an event if an unidentified person (who corresponds to actor in the above sample code) entered a restricted area or an unauthorized person entered the restricted area. The supporting facts for each of the patterns used in the rule are provided from different data models. RestrictedArea comes from the semantic space map; admitted comes dynamically from robot service rules; and EntranceLocation and actor come from the CIN. As such, the robot service knowledge performs the role of a service mediator as well as a data integrator, integrating all the situational data from the physical and virtual space and invoking services intelligently.

3.2. Dynamic Reconfiguration (DR). The DR module receives configuration change notifications originating from installation of a new device into the physical space, determines capabilities of the newly installed device by consulting a device profile repository, discovers the corresponding service components that can be deployed according to the new

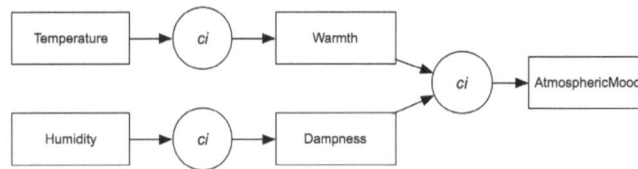

FIGURE 11. A sample CIN (Context Interpretation Network) for deriving atmospheric mood from temperature and humidity.

configuration, and finally deploys the service components. When a device is removed from the space, services associated with the removed device are also removed.

To describe the above scenario more specifically with respect to the u-Clips sensor network, when a new sensor node is introduced into the physical space, the node periodically announces a unique device identifier. Based on the identifier, the DR module tries to retrieve the corresponding device profile by accessing a device profile repository which contains pre-defined device profiles. We have implemented the device profile repository as a Java servlet-based web application, which accepts a device ID and responds with a corresponding device profile via HTTP protocol.

Once the DR module successfully retrieves the profile of the new device, it triggers a service discovery session to search for services that the new device provides. The new services are then deployed by adding them to the robot service knowledge or CIN of the CASE module. A distinct aspect in our design of DR is that all the services (e.g., robot service, context interpreter and the like) are implemented by ontology and rules rather than object codes or components. So, dynamic installation of a service can be realized simply by retrieving a text containing the specification of the service and saving it into a pre-determined directory. The CASE module in turn loads the service specification and executes the service.

An important issue in dynamic reconfiguration is the mechanism of service discovery. We adopted means-ends planning for discovering and composing services. The purpose of means-ends planning is to combine elementary tasks to achieve a given goal. An elementary task is modeled as an operation characterized by a quadraple (precondition, input, output, effect). Means-ends planning connects two operations if they have compatible inputs and outputs. That is, an operation o1 can be connected to an operation o2 if the set of o2's inputs is a valid subset of the set of o1's outputs. We modeled each sensor as an operation that has only outputs, and the robot service knowledge and the context interpretation nodes as operations having both inputs and outputs.

Fig. 12 shows sample models of operations. If a newly deployed sensor were identified as a `TemperatureSensor`, then its output, `Temperature`, is matched to the input of `WarmthInterpreter`, which makes the interpreter deployable. The models of operators are stored and managed by a service repository. In our system, the service repository is a simple web application with a query interface for service discovery and service retrieval.

3.3. Implementation of the Semantic Space. The functions of the semantic space—processing of world model, representation of robot service knowledge, and performing service discovery by means-ends planning—are all implemented by an embedded rule engine called eBossam [**34**]. eBossam, developed in C++ language, is a forward-chaining production rule engine built based on the RETE algorithm [**35**]. eBossam contains a set

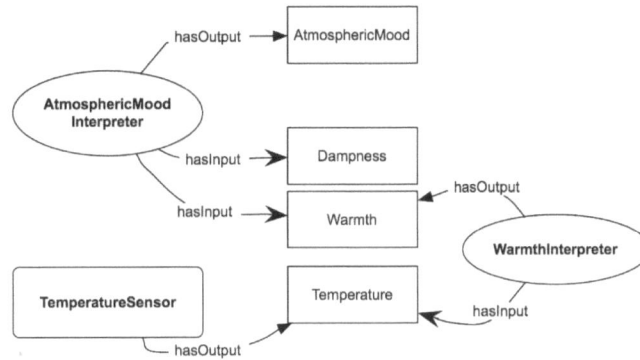

FIGURE 12. Sample models of operations in the semantic space.

of inference rules for performing simple reasoning over OWL ontology—e.g., deriving transitive closure from class hierarchy.

We implemented eBossam for embedded Linux operating system run on an ARM-based embedded processor. We have installed most of the semantic space functionalities inside a robot; the device profile repository and service repository, however, have been deployed as separate web applications to make the services publicly accessible. The semantic space utilizes a simple TCP sockets to interface with the physical space including a robot and the u-Clips sensor network.

4. The Virtual Space

The primary role of the virtual space is to provide users with a 2D or 3D virtual model of the physical space, thereby enabling the user to investigate and interact with the physical space in an intuitive way. Fusion of range and image data [36–40] and 3D reconstruction from a sequence of 2D images [41, 42] are two kinds of popular methods employed for modeling a 2D or 3D environment. In the proposed approach, SVG (Scalable Vector Graphics) and VRML (Virtual Reality Modeling Language) format are used for 2D and 3D map, respectively.

Construction of the virtual space is similar to building a robot simulator [43], where we implemented robot motion simulation, virtual experience and remote monitoring of the space through mobile devices. Hereinafter, we describe a procedure to construct a 2D and 3D model of the physical space.

Fig. 13 shows an experimental device for acquiring spatial snapshots. One laser scanner, two IEEE-1394 cameras, and an IR landmark-based localization sensor described in Section 2.1 comprise the data acquisition device. Each individual sensor is set up to be aligned along a vertical axis which goes through the geometric center of each sensor, thus simplifying data registration between range and image data. While the data acquisition device moves, the position and orientation of each data acquisition point is determined by the proposed localization sensor. Data are collected in such a way that a pair of successive data have overlapping parts.

Fig. 14 illustrates data flow of building 2D and 3D model for indoor environment. First, a 2D map is constructed from range data obtained from the laser scanner. Line features are extracted from a set of point clouds of range data. The coordinates of end points of each extracted line are used to generate a table (TMM: Table Metric Map in Fig.

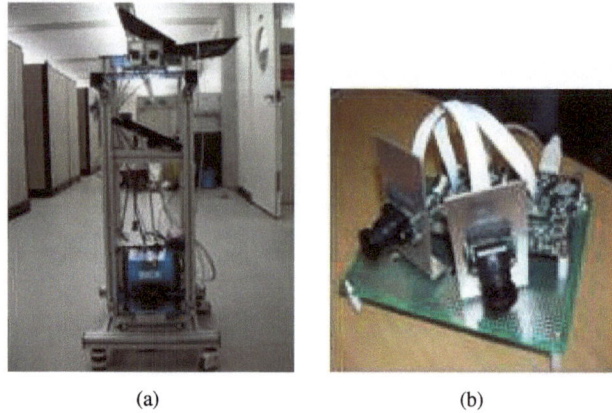

| (a) | (b) |

FIGURE 13. (a) An experimental device for acquiring spatial snapshots. The data acquisition device consists of an LMS200 laser scanner, a pair of IEEE-1394 Dragonfly cameras, and a precision localization sensor (see Section 2.1). (b) A close look of a pair of IEEE-1394 cameras.

14). A block metric map (BMM) is immediately obtained from the TMM. Finally, the 2D metric map is stored in the format of SVG for inter-operability between web-based applications.

Developing a 3D model relies on the already-made 2D map, consisting of a single floor plan and model patches representing walls. The developed 3D model is stored in the format of VRML. Fig. 15-(a) illustrates an example of a 3D model constructed from the 2D model of Fig. 14.

Texture data processing is carried out by using camera images. Image warping, stitching and cropping operations are applied to texture images. For texture image of the floor plan, a single image is obtained from the camera and the texture image is repeated to cover the entire floor plan. A sequence of texture images are collected and stitched together to form a wall image. For room-like structure, images are taken from four orthogonal directions and stitched together. Fig. 15-(b) shows a textured 3D model corresponding to Fig. 15-(a).

In order to model basic objects in the physical space, we apply the same procedure of modeling 3D space. One difference is addition of a laser scanner in vertical direction, which is intended to measure the height of each object in parallel with the laser scanner for measuring object geometry in horizontal direction as described above. Combining the scanned data with a texture image and utilizing geometric relationship between respective sensors and current location provided by the localization sensor, modeling of basic objects in the physical space are conducted. Fig. 16 illustrates an example of object modeling obtained from applying the sequence described above.

Time needed for gathering and processing data to make a 2D map of a 23.11 m × 25.52 m sized space (which is Fig. 15) is summarized in Table 1. In addition to the result of using a proposed localization sensor (see Section 2.1), we report the result when localization of the data acquisition device is carried out manually. It should be noted that the time needed for generating 2D and 3D maps depends on the size and structure of the space in question. The processing time includes geometric data processing for line extraction and texture

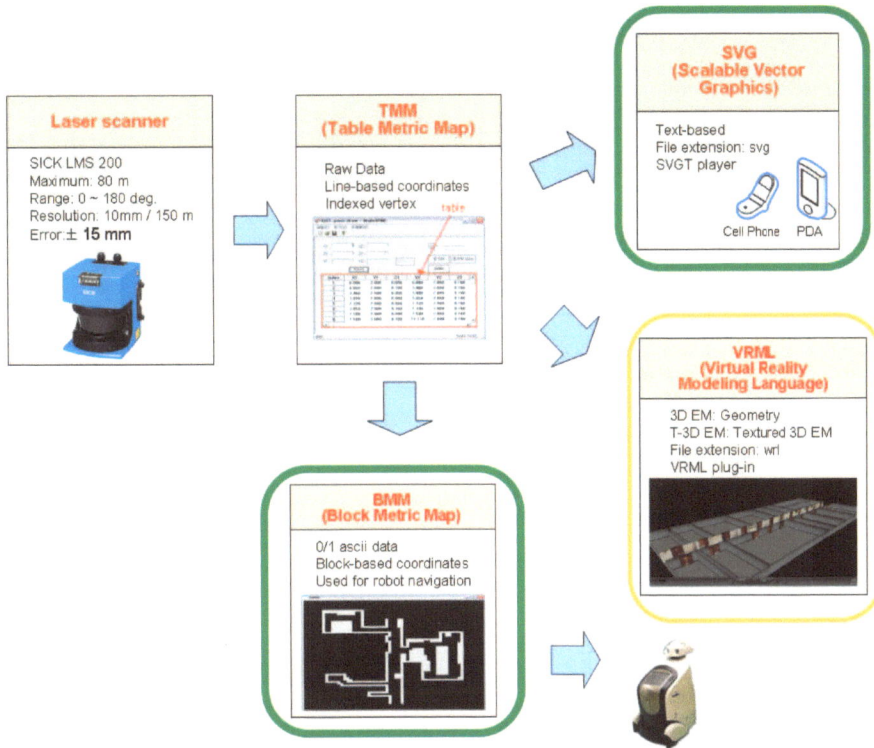

FIGURE 14. Data flow of building 2D and 3D model for indoor environment.

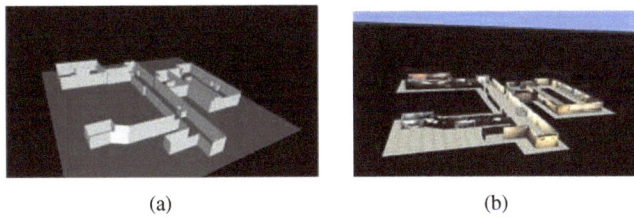

(a)   (b)

FIGURE 15. (a) 3D model generated from 2D model of Fig. 14. (b) The corresponding 3D texture model of (a).

FIGURE 16. An example of a 3D object model.

TABLE 1.　Time needed for data gathering and processing of a 3D Model in Fig. 15-(a).

Localization method	Data gathering (hr)	Data processing (hr)
StarLITE	0.5	0.5
Manual localization	2	0.5

stitching and cropping to the corresponding geometric data. After data processing is done, 2D and 3D map is automatically generated within a few seconds.

5. Implementation of the Proposed Ubiquitous Robotic Space

This section describes how the proposed URS can be applied to the real environment. We have developed a prototype robot security service directed to an office monitoring application. We have implemented the proposed URS in the ground floor of our building, the area of which was measured to be 23.11 m × 25.52 m.

5.1. Robot Security Application Utilizing the Proposed Ubiquitous Robotic Space.
The robot security application itself is not new as can be noticed from a large number of research works found in the literature. A brief overview of security robot systems is given in [44] including as early a development as reported in [45] and a more recent work describing a hand-off issue found in a networked robot for security application [46].

In general, a security robot is engaged in the tasks of detection of an abnormal situation and an appropriate reaction to the situation. The two tasks may involve various sub-tasks: zone-to-zone robot navigation crossing wireless network boundaries, remote perception of the environment, inference to understand the situation, generation of robot commands relevant to the situation and the like. Also, appropriate user interface to the space as well as the robot is necessary to provide the user with a means to control the whole system, which entails accurate modeling of the physical environment and visualization thereof.

The physical space comprises a wireless sensor network of Section 2.2, a localization network utilizing a precision localization sensor suite of Section 2.1, a mobile robot, and a URS server.

The localization network periodically provides location data to the robot and the URS server, with which the robot carries out a navigation task. At the same time, the user can check the current location of the robot by issuing an inquiry to the URS server.

We have developed a client program (see Fig. 17) for displaying the virtual space, environment data fed from the wireless sensor network, location information of the robot, and various status information of physical components (e.g., operation mode and battery level of the robot). The client program also provides a control interface for controlling the behavior of the robot in the case of manual operation of the URS.

The semantic space is responsible for monitoring the physical space and controlling the robot according to the given service scenario. An irregular situation such as intrusion or low battery level of the robot is detected from contextual information processing based on eBossam [34]; the semantic space issues a new navigation task to the robot upon detection of irregularity, where the irregular situation is reported to the user through CDMA communications network in the form of a short message service.

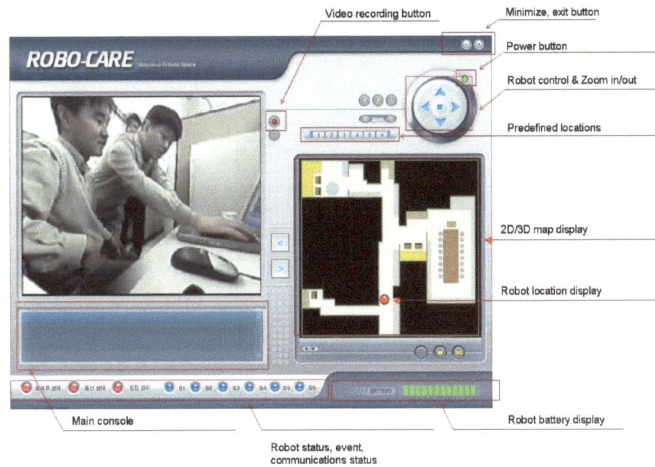

FIGURE 17. Implementation of ubiquitous robotic space at the client side.

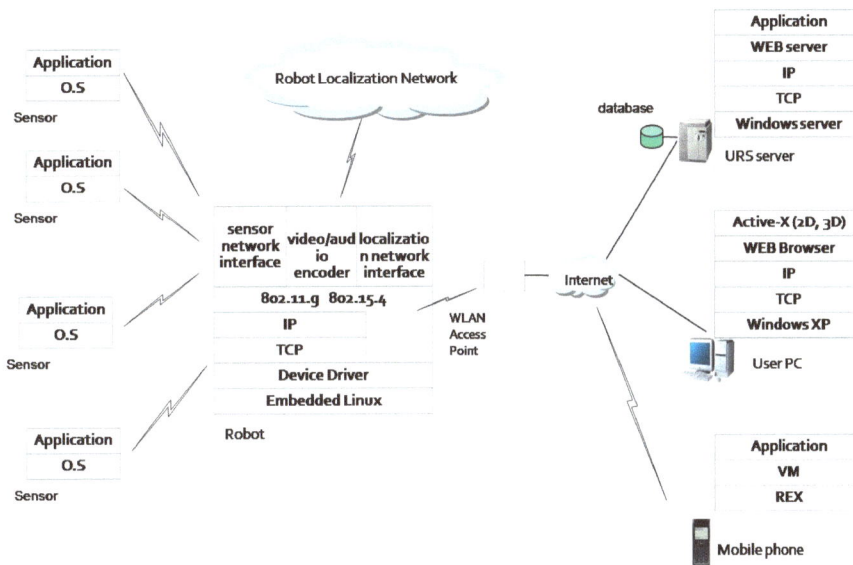

FIGURE 18. Component-wise view of the prototype robot security service.

Fig. 18 illustrates a component-wise view of the prototype robot security service. The application makes use of various networking and communications technologies such as ZigBee, WLAN, and WCDMA. Based on the wireless infrastructure, users can access a remote physical space and control the robot while watching video streams displayed on the screen of their terminals. A short video illustrating the above scenario can be obtained at http://robotask.etri.re.kr.

5.2. Data Communication Interface. We have defined data communication protocols to be used between the three spaces. As described earlier, we could develop the three

TABLE 2. Data type, length and description of each field of the employed message.

Field	Type	Length	Description
MSG ID	unsigned char	1 byte	message ID
opcode	unsigned char	1 byte	command code
data length	int	4 byte	length of a data area (measured in bytes)
data	data dependent	data dependent	data area

spaces independently of each other and integrate them easily by following the predefined protocols.

A message for exchanging data between the three spaces consists of six bytes of a message header and a data area of variable length. Table 2 illustrates data types and descriptions of respective fields of the message.

We have defined nine message IDs according to the nature of the corresponding data: UM_SERVER_MANAGE, UM_ROBOT_DATA, UM_ROBOT_CONTROL, UM_VIDEO, UM_SENSOR, UM_SERVICE, UM_MAP, UM_OBJECT, and UM_EVENT. For each message ID, we have defined various opcodes and accompanying data to specify operations against the corresponding data. UM_SERVICE, UM_MAP, and UM_OBJECT messages have been defined for download of service, download of environment map, and manipulation of objects respectively but are not used for the present implementation.

Table 3 presents the opcodes and corresponding data types used for transferring data between the three spaces. It should be noted that the messages in the table are only a partial set of messages, used to introduce the structure of the actual messages. For example, Table 3 provides only one typical example of UM_ROBOT_CONTROL, though UM_ROBOT_CONTROL is comprised of fourteen messages related to robot control, battery charging, and recovery of service rules.

5.2.1. **Data Transfer Between Physical Space and Semantic Space.** With respect to the robot security application, the physical space provides three kinds of messages: UM_ROBOT_DATA, UM_SENSOR, and UM_EVENT message.

The opcode of each message defines different data types and operation instructions. For example, a UM_ROBOT_DATA message with an opcode ROBOT_STATUS provides the semantic space with ID, battery level, and operation mode of the robot, where the operation mode includes idling, charging, and monitoring. Also, a UM_ROBOT_DATA message with an opcode ROBOT_POSITION periodically provides the position data of the robot for the semantic space.

The sink node installed inside the robot gathers sensor data from u-Clips sensor network and provides the sensor data for the semantic space in the form of a UM_SENSOR message. The data area of the message consists of sensor IDs and the corresponding sensor data of double type.

TABLE 3. Selected message definitions for transferring data inside the ubiquitous robotic space (P: Physical space, S: Semantic space, V: Virtual space, *s*: Source, *d*: Destination).

Message ID	Opcode	Data format	s	d	Description
UM_ROBOT _DATA	ROBOT_ID	BYTE robot_id	P	S	Transfer robot ID
	ROBOT _STATUS	BYTE robot_id, BYTE battery_level, BYTE task_mode	P	S, V	Transfer robot ID, battery level, and task mode upon status change or user's request
	ROBOT _POSITION	BYTE robot_id, BYTE zone_id, float x, y, theta	P	S, V	Periodically transfer current robot location
UM_SENSOR	SENSOR _DATA	BYTE sensor_id[16], double value	P	S	Transfer sensor data measured at u-Clips sensor nodes
UM_EVENT	EVENT _DETECT	BYTE event_type, BYTE sensor_id[16]	P	S	Transfer sensor ID and event type upon detection of an irregular situation (e.g., intrusion alarm)
UM_ROBOT _CONTROL	MOVE _PO-SITION	BYTE robot_id, BYTE zone_id, float x, y, theta, BOOL flag_rotate	S, V	P	Issues move command. flag_rotate determines whether or not to rotate after robot arrives at the designated position.
UM_SERVER _MANAGE	ROBOT _LIST	[ARRAY] BYTE robot_id, char robot_name, BOOL flag_connect	P	V	Transfer a list of connected robots
	USER_LIST	[ARRAY], char user_id[16], char passwd[16]	P	V	Transfer list of current users
UM_VIDEO	VIDEO _IN-FOR	BYTE robot_id, struct video_info	P	V	Transfer video information
	VIDEO _DATA	BYTE robot_id, video stream	P	V	Transfer video stream

The sink node also gathers motion sensor data, but treats the data as a UM_EVENT message with an opcode EVENT_DETECT different from the UM_SENSOR message.

Based on the above messages provided from the physical space, the semantic space carries out inference by using domain knowledge and task specifications. For example, the semantic space decides whether the message UM_EVENT represents an actual intrusion or a false alarm. Depending on whatever decision made, the semantic space issues a UM_ROBOT_CONTROL message to control the robot, the opcode of which specifies a basic behavior of the robot such as forward, backward motion, or rotation.

5.2.2. **Data Transfer Between Physical Space and Virtual Space.** As shown in Fig. 17, a client program is one embodiment of the virtual space.

Various status data are transferred from the physical space to the virtual space. Status data with an identifier UM_SERVER_MANAGE comprising a list of registered users and a list of robots connected to the server are transferred to the virtual space.

UM_ROBOT_DATA messages with opcodes of ROBOT_STATUS and ROBOT_POSITION are periodically transferred to the virtual space, the contents of which are displayed at the corresponding positions of the client program.

Video stream generated from a camera installed on top of the robot is transferred to the client program. UM_VIDEO messages with opcodes of VIDEO_INFO and VIDEO_DATA are used for transferring the video data to the client program.

Messages transferred from the virtual space to the physical space are mostly of user commands. Most of the messages with an identifier UM_SERVER_MANAGE are intended for administrative operations. Messages with an identifier UM_ROBOT_CONTROL are used for the user to manually control the robot; the same messages are used when the semantic space controls the robot according to the current situation. UM_VIDEO messages with different opcodes from the above (*e.g.*, VIDEO_START, VIDEO_STOP, VIDEO_CONTROL, etc) are also transferred from the virtual space to the physical space to operate a remote camera installed on top of the robot.

5.2.3. **Data Transfer Between Semantic Space and Virtual Space.** When a UM_EVENT message is generated with an opcode EVENT_DETECT, the semantic space carries out inference whether the message indicates an actual intrusion. If so, the semantic space transfers a UM_EVENT message with an opcode EVENT_ALARM to the virtual space notifying the user of the detection of intrusion.

The user generates various UM_ROBOT_CONTROL messages through the user interface of the client program, where the semantic space simply relays the generated messages to control the robot.

5.3. Comparison of the Proposed Ubiquitous Robotic Space with Previous Works.
In this section, we provide a brief overview about how the proposed ubiquitous robotic space architecture differs from the previous works considered to belong to the scope of our work.

Table 4 describes structural features of respective works introduced above and differences thereof from our work. The structure and functional roles of the three spaces are most similar to those described in [23–25]. Although the previous works envisioned an intelligent environment, actual environments are still far less intelligent to be characterized by ambient intelligence and networked resources are not dynamically configured yet to meet task specifications.

To summarize, the three spaces represent respective technology fields, which is not easy to integrate into a well-orchestrated system as described in [23]. By separating the respective spaces, we could develop each individual space with respect to its inherent features and requirements once we could define communication standards between the three spaces. In what follows, in addition to the robotic security service prototype, we describe our recent effort of implementing the proposed ubiquitous robotic space architecture in a large-scale artificial environment.

5.4. Application of the Proposed Ubiquitous Robotic Space to a Large-Scale Environment.
In this section, we describe our effort to apply the proposed architecture to a large-scale artificial environment. The environment has been recently built in the Incheon area of Korea. The primary purpose of this effort is to implement robotic services using robot navigation in a large-scale area crowded by people. The artificial building is supposed to represent a landmark of a future city (which is called *Tomorrow City*) characterized by the concept of ambient intelligence, incorporating the state-of-the-art IT infrastructure as well as robotics technology. The overall size of the building amounts to $31,670\ m^2$, where 19 mobile robots of five different services are involved. Figure 19-(a) and (b) show the inside and outside view of the building, respectively.

Four mobile robots operate in the Sunken plaza, where robot localization is carried out by utilizing a DGPS system of 20 cm accuracy, wheel encoders, and special-purpose barcodes sparsely attached on the surface of the plaza. Inside the building, four different

TABLE 4. Description of previous works and differences from the proposed architecture.

Name	Purpose	Key elements	Features	Difference from proposed URS
-	To control networked robotic systems [21]	Wireless network, networked robot	Networked robots controlled by a remote robotic server	A primitive hardware configuration for a networked robot system is provided. No description corresponding to semantic or virtual spaces is given.
Intelligent Space [8]	To transform an ordinary space into an intelligent space	Distributed Intelligent Networked Device (DIND)	Smart sensor nodes connected and controlled through a wireless network	A sensing network based on DIND is provided, i.e., a network of smart cameras. No consideration corresponding to semantic or virtual spaces is given.
Networked Robotic Framework [22]	To integrate embedded systems with sensor and actuator networks into multi-robot and distributed agent architecture	Module pool and seamless integration of resources distributed across different fixed and mobile robot platforms	Dynamic reconfigurable architecture utilizing module pool	The role of a task module is similar to that of the proposed semantic space, but no consideration is given to the role of the proposed virtual space. Utilization of sensor network and localization infrastructure is not explicitly described.
Physically Embedded Intelligent System (PEIS) [23–25]	To integrate robot systems into smart environment characterized by ambient intelligence	PEIS component, PEIS kernel	Dynamic configuration of modeling, deliberation, perception, and control blocks distributed across PEIS-ecology	The higher layer consisting of modeling and deliberation blocks corresponds to the proposed semantic space, and the lower layer of perception and control blocks corresponds to the proposed physical space; no consideration to the virtual space is given. Each building block is implemented using different technologies—for example, world modeling is carried out in LOOM language, whereas in the proposed architecture, OWL Web ontology language is used.
Ubiquitous Function (UF) Service [27]	To control an intelligent space comprised of networked robotic systems and ambient intelligence	UF smart logic service, UF smart object service, and UF smart discovery service	Similar to [22], comprised of publishing, finding, and binding of UF smart objects and smart logic services	Description related to the proposed virtual space is not given. Web service technology was used to implement the respective services, whereas each space of the proposed URS is implemented independent of each other by utilizing the technology most relevant to the space property.

types of robots provide services of visitor guidance, baggage delivery, patrol along the hallway of a mall, and beverage delivery in a shop upon the order of guests.

A remote URS server handles various tasks such as maintaining connectivity of the mobile robots to the Internet, application services according to the separate robot tasks, and remote control of the individual robots utilizing 2-D and 3-D maps developed. The

(a)

(b)

FIGURE 19. Large-scale robot navigation based on the proposed ubiquitous robotic space architecture. (a) Two mobile robots are navigating inside the so-called Sunken plaza. (b) The outside view of the tomorrow city.

internal architecture of the robotic services are actually the same as that introduced so far, only differing in the service knowledge base and some additions of specific service rules.

The entire area is full of wireless LAN infrastructure and a roaming function is automatically carried out as soon as a robot crosses the boundary of a current WLAN access point. By doing so, the robot could maintain seamless data communication with the URS server, thereby achieving smooth navigational transition between different AP regions without noticeable motion discontinuity.

For mapping of the large-scale area, we have developed a portable mapping device. Figure 20-(a) shows the mapping device developed. The introduction of the mapping device originated from the complicated internal structure of the building, which entailed a difficult decision to make for a mobile robot exploring an unknown environment. Moreover, while in construction, the robot had to detect and avoid a variety of construction materials scattered on the ground during the mapping process, making a theoretically rigid

(a)   (b)

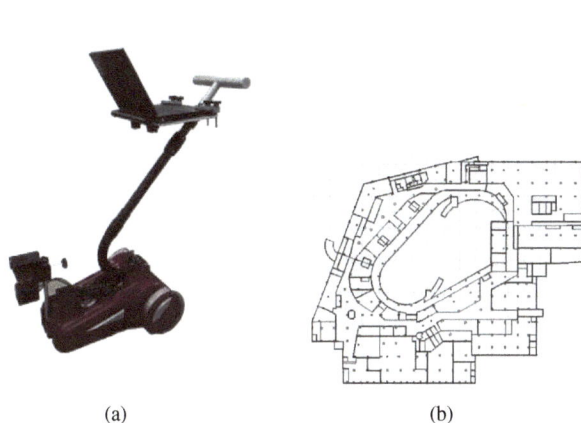

FIGURE 20. (a) A portable device developed for large-scale mapping. The device carries a laser scanner for extracting the structure of the environment, wheel encoders, and a DSP board for data association between the laser scanner data and the encoder readings. (b) 2D-map of the *Tomorrow City*.

SLAM process impractical. We also developed a software tool to edit the initial mapping data to generate a clean, accurate map.

The whole sequence of the mapping process starts with conversion of a CAD file of the building to the corresponding grid map. After the conversion, we mapped hundreds of landmarks consisting of special barcodes attached on the ceiling and the floor of the building. Finally, we built a 2D and 3D object map (see Fig. 20-(b)) on top of the landmark map, thereby enabling a remote operator to conveniently monitor the current locations of the robots and to control the robots.

Developing a large-scale robot navigation application requires many different research and development teams coming from different organizations. The entire robotic service system, however, could be built by utilizing the proposed ubiquitous robotic space without noticeable modifications of the proposed framework. Most of the works has been dedicated to build the maps for robot operation, service scenario development, and adjusting each individual robot platform to operate well instead of updating the proposed architecture. The whole robotic service system was in full service as one of five major theme spaces during so-called *Incheon Global Fair and Festival* which lasted from Aug. 8 to Oct. 25.

6. Conclusion

In this paper, we have proposed a ubiquitous robotic space where a robot or a human can have information assistance from the space and a specific robotic task can be carried out efficiently. As a demonstration of the proposed URS, we have implemented a prototype robot security application.

The number and characteristics of employed physical components and complexity entailed to interconnect the three spaces may vary according to a particular security scenario. The proposed architecture, however, is capable of accommodating wide range of service requirements since it can be thought of as an instance of hybrid control architecture; the semantic space resembles a deliberative layer implementing task intelligence and domain knowledge of a given service scenario while more reactive tasks are handled within the physical space.

Also, each space can be developed independently of each other under the condition that interfaces among the three spaces are well defined, which is a useful feature of the proposed URS for integrating heterogeneous components of the three spaces.

One of our future works includes integrating a camera network to the developed robotic service system for monitoring human activities, thereby carrying out an experimentation about how a robotic system can effectively support human daily life. Another future goal is to establish a standardized robotic service platform, with which a variety of separate developers and users can easily develop and utilize the proposed robotic service framework.

Acknowledgement

This work was supported by the R&D program of the Korea Ministry of Knowledge and Economy (MKE) and the Korea Evaluation Institute of Industrial Technology (KEIT). [2005-S-092-02, USN-based Ubiquitous Robotic Space Technology Development; 2008-S-031-01, Hybrid u-Robot Service System Technology Development for u-City]

Bibliography

[1] W. Yu, J. Lee, Y. Ha, M. Jang, J. Sohn, Y. Kwon, and H. Ahn, "Design and implementation of a ubiquitous robotic space," *IEEE Trans. Automation Science and Engineering*, vol. 6, no. 4, pp. 633–640, Oct. 2009.

[2] IFR Statistical Department, *World Robotics 2007.*   Frankfurt, Germany: VDMA Robotics & Automation, 2007.

[3] M. Weiser, "The computer for the 21st century," *Scientific American*, vol. 265, no. 3, pp. 66–75, 1991.

[4] W. Weber, J. M. Rabaey, and E. Aarts, Eds., *Ambient Intelligence.*   Springer, 2005.

[5] D. J. Cook, M. Youngblood, E. Heierman, K. Gopalratnam, S. Rao, A. Litvin, and F. Khawaja, "Mavhome: An agent-based smart home," in *IEEE Int. Conf. Pervasive Computing and Communications*, Mar. 2003, pp. 521–524.

[6] T. Kirste, *True Visions: The Emergence of Ambient Intelligence.*   Berlin: Springer, 2006, ch. 17, pp. 321–337.

[7] J. Lee and H. Hashimoto, "Controlling mobile robots in distributed intelligent sensor network," *IEEE/ASME Trans. Mechatronics*, vol. 50, no. 5, pp. 890–902, Oct. 2003.

[8] J. Lee, K. Morioka, N. Ando, and H. Hashimoto, "Cooperation of distributed intelligent sensors in intelligent environment," *IEEE/ASME Trans. Mechatronics*, vol. 9, no. 3, pp. 535–543, Sep. 2004.

[9] T. Sato, "Robotic room: Human behavior measurement, behavior accumulation and personal/behavioral adaptation by intelligent environment," in *IEEE/ASME Int. Conf. Advanced Intelligent Mechatronics*, Jul. 2003, pp. 20–24.

[10] T. Sato, T. Harada, and T. Mori, "Environment-type robot system "Robotic Room" featured by behavior media, behavior contents, and behavior adaptation," *IEEE/ASME Trans. Mechatronics*, vol. 9, no. 3, pp. 529–534, Sep. 2004.

[11] N. Y. Chong, H. Hongu, K. Ohba, S. Hirai, and K. Tanie, "A distributed knowledge network for real world robot applications," in *Proc. IEEE/RSJ Int. Conf. on Intelligent Robots and Systems (IROS)*, Sendai, Japan, Sep. 2004, pp. 187–192.

[12] S. Sugano and Y. Shirai, "Robot design and environment design: Waseda robot-house project," in *SICE-ICASE Int. Joint Conf*, Oct. 2006, pp. 31–34.

[13] J. Borenstein, H. R. Everett, and L. Feng, "Where am I? sensors and methods for mobile robot positioning," Univ. of Michigan, Tech. Rep., Apr. 1996.

[14] R. C. et al., "Hidden issues in deploying an indoor location system," *IEEE Pervasive Computing*, vol. 6, no. 2, pp. 62–69, Apr.-Jun. 2007.

[15] J. Hightower and G. Borriello, "Location systems for ubiquitous computing," *IEEE Computer*, vol. 34, no. 8, pp. 57–66, Aug. 2001.

[16] K. Pahlavan, X. Li, and J. Makeka, "Indoor geolocation science and technology," *IEEE Communications Magazine*, pp. 112–118, Feb. 2002.

[17] T. S. Rappaport, J. H. Reed, and B. D. Woerner, "Position location using wireless communications on highways of the future," *IEEE Communications Magazine*, pp. 33–41, Oct. 1996.

[18] G. Alankus, N. Atay, C. Lu, and O. B. Bayazit, "Adaptive embedded roadmaps for sensor networks," in *IEEE Int. Conf. Robotics and Automation*, Apr. 2007, pp. 3645–3652.

[19] M. A. Batalin, G. S. Sukhatme, and M. Hattig, "Mobile robot navigation using a sensor network," in *IEEE Int. Conf. Robotics and Automation*, Apr. 2004, pp. 636–642.

[20] M. A. Batalin and G. S. Sukhatme, "Using a sensor network for distributed multi-robot task allocation," in *IEEE Int. Conf. Robotics and Automation*, Apr. 2004, pp. 158–164.

[21] R. C. Luo, K. L. Su, S. H. Shen, and K. H. Tsai, "Networked intelligent robots through the Internet: issues and opportunities," *Proc. IEEE*, vol. 91, no. 3, pp. 371–382, Mar. 2003.

[22] D. I. Baker, G. T. McKee, and P. S. Schenker, "Network robotics, a framework for dynamic distributed architectures," in *IEEE/RSJ Int. Conf. Intelligent Robots and Systems*, Sep. 2004, pp. 1768–1771.

[23] R. Lundh, L. Karlsson, and A. Saffiotti, "Plan-based configuration of an ecology of robots," in *IEEE Int. Conf. Robotics and Automation*, Apr. 2007, pp. 64–70.

[24] M. Broxvall, M. Gritti, A. Saffiotti, B. Seo, and Y. Cho, "PEIS ecology: integrating robots into smart environments," in *IEEE Int. Conf. Robotics and Automation*, May 2006, pp. 212–218.

[25] A. Saffiotti and M. Broxvall, "Ecologies of physically embedded intelligent systems," Dept. Technology, Örebro Univ., Tech. Rep., Dec. 2004.

[26] R. R. Murphy, *Introduction to AI Robotics*. The MIT Press, 2000, ch. 7.

[27] B. Kim, N. Tomokuni, K. Ohara, T. Tanikawa, K. Ohba, and S. Hirai, "Ubiquitous localization and mapping for robots with Ambient Intelligence," in *IEEE/RSJ Int. Conf. Intelligent Robots and Systems*, Oct. 2006, pp. 4809–4814.

[28] H. Ahn and W. Yu, "Environmental-adaptive RSSI-based indoor localization," *IEEE Trans. Automation Science and Engineering*, vol. 6, no. 4, pp. 626–633, Oct. 2009.

[29] W. Lin, S. Jia, T. Abe, and K. Takase, "Localization of mobile robot based on ID tag and WEB camera," in *Proc. IEEE Int. Conf. on Robotics, Automation and Mechatronics*, Singapore, Dec. 2004, pp. 851–856.

[30] Y. Nagumo and A. Ohyae, "Human following behavior of an autonomous mobile robot using light-emitting device," in *Proc. IEEE Int. Workshop on Robot and Human Interactive Communication*, Bordeaux-Paris, Sep. 2001, pp. 225–230.

[31] J. S. Park and M. J. Chung, "Path planning with uncalibrated stereo rig for image-based visual servoing under large pose discrepancy," *IEEE Trans. Robotics and Automation*, vol. 19, no. 2, pp. 250–258, Apr. 2003.

[32] *ZigBee Network Layer Specification 1.0*, ZigBee Alliance, Dec. 2004.

[33] *Standards 802.15.4: Wireless Medium Access Control (MAC) and Physical Layer (PHY) Specifications for Low-Rate Wireless Personal Area Networks (LR-WPANs)*, IEEE Computer Society, 2003.

[34] M. Jang and J. Sohn, "Bossam: An extended rule engine for OWL inferencing," in *Proc. RuleML (LNCS Vol. 3323)*, Nov. 2004.

[35] C. Forgy, "RETE: A fast algorithm for the many pattern/many object pattern match problem," *Artificial Intelligence*, no. 19, pp. 17–37, 1982.

[36] D. Hähnel, W. Burgard, and S. Thrun, "Learning compact 3D models of indoor and outdoor environments with a mobile robot," *Robotics and Autonomous Systems*, vol. 44, no. 1, pp. 15–27, Jul. 2003.

[37] O. Wulf, K. O. Arras, H. I. Christensen, and B. Wagner, "2D mapping of cluttered indoor environments by means of 3D perception," in *IEEE Int. Conf. Robotics and Automation*, Apr. 2004, pp. 4204–4209.

[38] P. Biber, H. Andreasson, T. Duckett, and A. Schilling, "3D modeling of indoor environments by a mobile robot with a laser scanner and panoramic camera," in *IEEE/RSJ Int. Conf. Intelligent Robots and Systems*, Sep.-Oct. 2004, pp. 3430–3435.

[39] P. Dias, V. Sequeira, F. Vaz, and J. G. M. Goncalves, "Registration and fusion of intensity and range data for 3D modeling of real world scenes," in *Int. Conf. 3-D Digital Imaging and Modeling*, 2003, pp. 418–425.

[40] I. Stamos, L. C. Chen, G. Wolberg, G. Yu, and S. Zokai, "Integrating automated range registraion with multiview geometry for the photorealistic modeling of large-scale scenes," *Int. J. Computer Vision (Special Issue)*, vol. 78, no. 2-3, pp. 237–260, Jul. 2008.

[41] O. D. Faugeras, *Three-Dimensional Computer Vision: A Geometric Viewpoint*. Cambridge, MA: MIT Press, 1993.

[42] Y. Tan, J. Hua, and M. Dong, "3D reconstruction from 2D images with hierarchical continuous simplices," *The Visual Computer: Int. J. Computer Graphics*, vol. 23, no. 9, pp. 905–914, Aug. 2007.

[43] J. Faust, C. Simon, and W. D. Smart, "A video game-based mobile robot simulation environment," in *IEEE/RSJ Int. Conf. Intelligent Robots and Systems*, Oct. 2006, pp. 3749–3754.

[44] W. Yu, J. Lee, H. Chae, K. Han, Y. Lee, and Y. Ha, "Robot task control utilizing human-in-the-loop perception," in *IEEE Int. Symposium on Robot and Human Interactive Communication (RO-MAN)*, Aug. 2008, pp. 395–400.

[45] S. Y. Harmon, "The ground surveillance robot (GSR): An autonomous vehicle designed to transit unknown terrain," *IEEE J. Robotics and Automation*, vol. 3, no. 3, pp. 266–279, Jun. 1987.

[46] C. Ku and Y. Cheng, "Remote surveillance by network robot using WLAN and mobile IPv6 techniques," in *Proc. TENCON*, 2007, pp. 1–4.

CHAPTER 10

Toward High-Performance Stable Haptic Teleoperation over the Internet: Passive Set-Position Modulation (PSPM) Approach

Dongjun Lee and Ke Huang

Department of Mechanical, Aerospace, and Biomedical Engineering

University of Tennessee

Knoxville, TN 37996, USA

ABSTRACT. This book chapter proposes the recently-developed passive set-position modulation (PSPM) framework as a promising means toward passive, yet, high-performance haptic teleoperation over the Internet with significant varying-delay and data-loss. We first briefly review some representative control techniques in haptics and time-delay teleoperation (i.e. virtual coupling, passivity observer/controller (or PO/PC), scattering/wave, and proportional-derivative (PD) based approaches) to provide better contextuation/justification for the development of PSPM, with a particular emphasis on how to improve performance. We then provide a condensed derivation of the PSPM framework and a summary of key properties of the PSPM framework and the PSPM based Internet haptic teleoperation, along with the complete algorithm of the PSPM and some illustrative experimental results. Some comments on the future research directions are also given.

1. Motivation

This book chapter is concerned with the problem of haptic teleoperation between two robotic systems networked over the ubiquitous yet unreliable Internet communication. Here, the term "haptic" is loosely used to describe the teleoperation problems, where the master and slave robotic systems mechanically interact with their human users or environments via their respective power port (i.e. force F_i and velocity $\dot{q}_i$ pair - see Fig. 1). Of course, being teleoperation implies that such mechanical couplings at the master and the slave systems should be related to each other. This needs to be done, here, via the Internet communication, which is often characterized by its (infamous) communication delay and packet loss, both of them evolving in a unpredictable or random/stochastic fashion and having tendency to easily destabilize the total closed-loop system.

Many powerful applications have spurred the theoretical/technological advancements of the haptic teleoperation: tele-surgery, remote civil structure construction, tele-driving [5], and micro-assembly with motion scaling, to name a few. Some adjacent areas/problems, that necessitate networking of robots, humans, and their real/virtual environments, have also been benefited by the gained insights and results of teleoperation, including multi-user networked haptic interaction [6,7], remote haptics [3], and even multi-agent consensus [8]. The impact of haptic Internet teleoperation will be even greater, since it will then contribute to and/or enable us to achieve those networking applications over the pervasive and affordable Internet.

Yet, this Internet haptic teleoperation problem is challenging. This is particularly due to its two aspects, that make even merely maintaining (interaction) stability a non-trivial task: 1) human dynamics often un-modeled, uncertain and/or too complicated to be modeled as a mathematical tractable model; and 2) randomly-varying communication delay and packet loss, well known for their ability to destabilize any otherwise stable control systems. The concept of passivity has been pursued to address this interaction stability problem, since, by enforcing passivity of the closed-loop teleoperation system (including the Internet communication), we can guarantee its interaction stability with *any* passive human operators and environments. As pointed out in [9], passive human assumption is generally not so a conservative one in practice and also the environment, in which the task is performed, is often composed of just passive mechanical components (e.g. mass-spring-damper type environments).

This idea of passivity has provided theoretically sound frameworks for many works for passive (or robustly-stable) teleoperation. Since the seminal work of [11], that recasts the delayed communication channels into the power transmission lines and relates

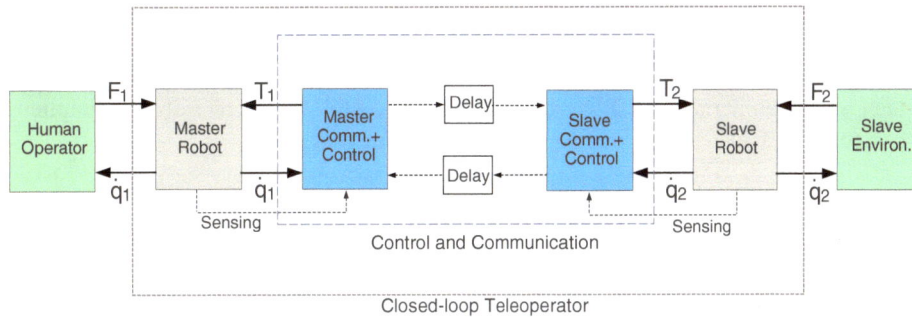

FIGURE 1. Teleoperation systems consisting of master and slave robots (from [**10**]). For the Internet communication, only the delays are shown here for simplicity.

interaction stability of the teleoperation system to the bounded realness [**12**] of the communication scattering matrix (with remaining part of the system composed of only passive components), the scattering-based approach and its later wave-variable reformulation [**13–15**] have been virtually the only way to achieve passive teleoperation with constant delay. Then, a new result was proposed in [**10**], where, using the concept of controller passivity and Parseval's identity, it was shown that the surprisingly simple linear PD (proportional-derivative) control (i.e. spring-damper) can also enforce closed-loop passivity of the teleoperation with constant delay. Some subsequent refinements [**16, 17**] ensue this PD scheme.

Although ensuring interaction stability (or passivity) is challenging and has a theoretical merit by itself, it is not the whole story about the haptic teleoperation. The teleoperation system should also provide a reasonable level of performance, that is, tele-presence of the human users into the slave environment, or, at least, utilizable perception of the slave environment. Unfortunately, due to their conservatism for robust stability, the aforementioned passivity-based teleoperation schemes are often associated with poor performance (e.g. position drift in the presence of packet loss for the scattering-based teleoperation; large local damping in the PD-based teleoperation in free motion, particularly with long communication delay). Similar performance issue is also relevant to the twin problem of teleoperation - haptic interface control problem, where passivity is again desired for interaction stability with a wide range of human operators, while some level of performance (e.g. z-width [**18**]) is necessary for providing the intended virtual realism.

In the next Sections 2 and 3, we will briefly review some representative control techniques for haptics and delayed teleoperation to provide better contextuation/justification for the development of the passive set-position modulation (PSPM) framework, with a particular emphasis on how to achieve passive, yet, high-performance Internet haptic teleoperation. A complete algorithm and a summary of key properties of the PSPM framework will then be explained in Sec. 4.2 along with some illustrative experimental results. For more detailed derivations/developments/proofs of the PSPM, we refer readers to [**1–4**].

2. Haptic Interface Control

As said above, the problem of haptic interface control can be considered as a twin of that of the teleoperation, given that both of them often require closed-loop passivity for interaction stability with a wide range of human users and physical/virtual environment, although each has its own indigenous issues (e.g. visibility and uncertainty of slave environment for teleoperation; mechanics simulation and graphics rendering for haptics). One of the most frequently used haptic interface control frameworks is **virtual coupling** [19], which is very similar to the PD-based teleoperation control. More precisely, for simplicity, let us consider 1-DOF haptic device with the following dynamics:

$$m\ddot{x} + b_e\dot{x} = \tau + f$$

where $x \in \Re$ is the position, $m > 0$ is the mass, $b_e > 0$ is the (continuous-time) device damping, and $\tau, f \in \Re$ are the control and human force. Then, the virtual coupling control is given by: with the zero-order-hold (ZOH), for $t \in [t_k, t_{k+1})$,

$$\tau(t) = -b\frac{x_k - x_{k-1}}{\Delta T} - k(x_k - y_k)$$

where $b, k > 0$ are the damping and spring gains, $x_k \in \Re$ is the sampled position x at t_k, ΔT is the sampling rate (typically in the order of several milliseconds), and $y_k \in \Re$ is the position of the the haptic device's presence in the virtual environment (or virtual proxy). Here, note the backward numerical differentiation for the damping control velocity (i.e. $(x_k - x_{k-1})\Delta T$).

The performance of this virtual coupling control is then characterized by the z-width [18], i.e., given (T, b, b_e, m), what is the largest possible value of k without violating the closed-loop passivity: for all $T \geq 0$, $\exists$ a bounded constant $d \in \Re$ s.t.

$$(1) \qquad\qquad \int_0^T f\dot{x}dt \geq -d^2$$

with the above virtual coupling control $\tau(t)$. Of course, if there is no sampling and zero-order-hold (ZOH) in the above $\tau(t)$, the virtual coupling τ will consist of only intrinsically passive continuous-time damping and spring, thus, regardless of how large b, k are, the closed-loop haptic interface will satisfy the above closed-loop passivity (with discrete passive virtual environment of y_k). This is not true with the above $\tau(t)$, since some energy is inevitably generated via the sampling and ZOH processes (i.e. energy leak [20]). Yet, still, if the continuous-time device damping b_e is large enough, this energy generation can be absorbed and the closed-loop passivity can be still ensured. In particular, we have the following celebrated Colgate's passivity condition (from virtual wall proof [21] - see [22] for time-domain proof for virtual coupling):

$$(2) \qquad\qquad b_e \geq \frac{k\Delta T}{2} + b$$

which shows that, given the sampling rate ΔT, there is a upper-bound of the z-width. Moreover, the virtual discrete damping b is not helpful at all to extend the z-width. Only the real continuous device damping b_e can do so.

For a typical haptic application with a typical haptic device (i.e. k, b_e specified), it is usually required to have ΔT less than 1ms (i.e. sampling faster than 1kHz). This is possible for a simple virtual environment (e.g. 1-degree-of-freedom (DOF) virtual wall). Yet, it is not generally so for a complicated virtual environment. Even more, most of interesting virtual environments are complex, requiring a longer ΔT for more computation. We cannot simply increase the device damping b_e either, since it is the real physical damping, thus, not easily adjustable via software. Also, a larger b_e will result in sluggish system response, potentially detrimental for the intended virtual realism, particularly in free motion. In other words, the passivity condition (2) is too restrictive to achieve high performance for many interesting haptic virtual environments.

Now, suppose that we set the virtual coupling gains so that the passivity condition (2) is violated. Then, if we are measuring the closed-loop passivity (18) (i.e. using force/velocity sensing), we could often observe that the closed-loop passivity (18) is violated not all the time, but for some (no so frequent) time instances. This implies that the passivity condition (2) may be potentially too conservative, as the virtual coupling $\tau(t)$, which is essentially a time-invariant constant gain state feedback, blindly enforces passivity for all the time and for all the frequency range, with its design being tuned for the (rarely-happening) "worst-case" scenario. This conservatism, therefore, may be reduced by adjusting the structure/gain of the virtual coupling τ from time to time depending on the system's passivity behavior.[1] This we believe is one of the key aspects behind the wide-usage of the **PO/PC** (passivity-observer/passivity-controller) framework [**23**] for better performance haptic interface control.

The (series impedance causality) PO/PC may be roughly summarized as follows. First, assume that the system behavior can be described well enough in discrete time. Define the haptic interface control at t_k s.t.: during $[t_k, t_{k+1})$,

$$\tau_k := \tau_k' + b_k v_k$$

where τ_k' is the unmodulated control received from the virtual environment at t_k (e.g. the above virtual coupling), v_k is the sampled and ZOH of $\dot{x}(t)$, τ_k will be the ZOH control for the haptic interface, and b_k is the variable damping to compensate for any shortage of passivity as defined below. Also, at the onset of t_n, we can compute the energy to be generated by τ_n, if we set $b_n = 0$, s.t.:

$$E_{\text{obsv}}(n) = \sum_{k=0}^{n-1} \tau_k v_k \Delta T + \tau_n' v_n \Delta T$$

where n is discrete time index. We want this energy generation to be bounded (e.g. less than zero), since, from the open-loop passivity of the haptic device and the notion of controller passivity [**24**], it will then in turn imply the closed-loop passivity (18). This bookkeeping of $E_{\text{obsv}}(n)$ constitutes PO. If $E_{\text{obsv}}(n) > 0$ for any n, passivity is violated and some action is needed to correct it. That is when the PC comes in with the damping injection b_n designed s.t. at t_n, if $E_{\text{obsv}}(n) > 0$,

$$b_n = \frac{E_{\text{obsv}}(n)}{v_n^2 \Delta T}$$

or $b_n = 0$, otherwise. This damping injection will then absorb the (surrogated) non-passive energy generation by τ_n', thereby, enforcing passivity.

Even if this PO/PC framework is without shortcomings (e.g. neglecting inter-sample continuous-time dynamics; unbounded control action and noisy behavior due to v_n^2 in the denominator of b_n [**25, 26**]), its wide acceptance for better performance haptic interface control suggests us that, perhaps, for the Internet haptic teleoperation, rather than constant gain state feedback type controls (e.g. scattering and PD schemes), control laws with some selectively varying structures (similar to the time-varying damping b_k for the PO/PC - not to be confused with variable structure control) may be better suited to achieve high performance and passivity. This is exactly the idea behind the development of our PSPM framework. Before introducing the PSPM, to make a better context with the state of the

---

[1]See also [**9**] for using a more sophisticated and numerical control law synthesis than the above virtual coupling $\tau(t)$ to selectively enforce either passivity or stability for different operational frequency ranges.

art, let us first briefly review the two main frameworks for the passive Internet haptic tele-operation: scattering-based and PD-based approaches.

3. Haptic Teleoperation with Delay

Typically, master and slave robots are modeled by the following multi-DOF nonlinear Lagrangian dynamics:

$$(3) \qquad M_1(x_1)\ddot{x}_1 + C_1(x_1,\dot{x}_1)\dot{x}_1 = \tau_1 + f_1$$

$$(4) \qquad M_2(x_2)\ddot{x}_2 + C_2(x_2,\dot{x}_2)\dot{x}_2 = \tau_2 + f_2$$

where $x_\star$ is the configuration (or, generalized coordinate: e.g. for 6-DOF robot, (x,y,z)-translation and pitch/yaw/role angles), $M_\star$ is the symmetric/positive-definite inertia matrix, $C_\star$ is the Coriolis matrix, $g_\star$ is the gravitational term, and $\tau_\star, f_\star$ are the control action (e.g. given by actuator motors) and the external force (e.g. human force, or contact force). Here, we assume that the gravitational terms for the master and slave systems have been locally canceled out.

As shown in Fig. 1, the closed-loop teleoperation system can then be thought of as a two port system, each port mechanically interacting with the human operator and the slave environment. Then, we can guarantee interaction stability with *any* passive human users and environments by rendering the closed-loop teleoperation system to be two-port passive: $\exists$ a bounded constant $d \in \Re$ s.t.

$$\int_0^T [f_1^T \dot{x}_1 + f_2^T \dot{x}_2]dt \geq -d^2$$

for all $T \geq 0$. To check this closed-loop passivity in general requires to consider the closed-loop dynamics of the teleoperation system, which is often complicated and nonlinear. Instead, the following two-port controller passivity can be used [24]: $\exists$ a bounded constant $c \in \Re$ s.t.

$$\int_0^T [\tau_1^T \dot{x}_1 + \tau_2^T \dot{x}_2]dt \leq c^2$$

for all $T \geq 0$. Then, as shown in [10, 24], due to the open-loop passivity (i.e. energy conservation) of the master and slave Lagrangian dynamics (3)-(4), this controller passivity implies the closed-loop passivity. Furthermore, checking of this controller passivity is much easier than that of the closed-loop passivity, since the control laws $\tau_\star$ with $\dot{x}_\star, x_\star$ as their input have usually much simpler structure than that of the closed-loop dynamics (e.g. linear time-invariant structure of the PD control [10]). In other words, this controller passivity allows us to reach the closed-loop passivity while bypassing the nonlinearity of the robotic dynamics.

Now, let us consider the case where the master and slave communication undergoes constant delay. For this case, the **scattering-based framework** [11] and its wave-variable reformulation [13–15] has been the dominating scheme to achieve passive teleoperation. See Fig. 2. The key idea of this, as proposed in [11], is to relate the passivity to the bounded realness (BR) of the scattering matrix. More precisely, let us define $\tau_d := [\tau_1^d, \tau_2^d]^T$, $v_d := [v_1^d, v_2^d]^T$, where $\tau_\star^d$ and $v_\star^d$ are respectively the control output from and the desired velocity input to the master and slave local control modules. Here, the control $\tau_\star^d$ aims to coordinate $\dot{x}_\star$ and v^d (e.g. PI-control or coordinating torque [11]) with the relation $\tau_\star^d = -\tau_\star$. For simplicity, here, let us only consider the scalar case with $\tau_\star^d, v_\star^d \in \Re$. Then, in the Laplace

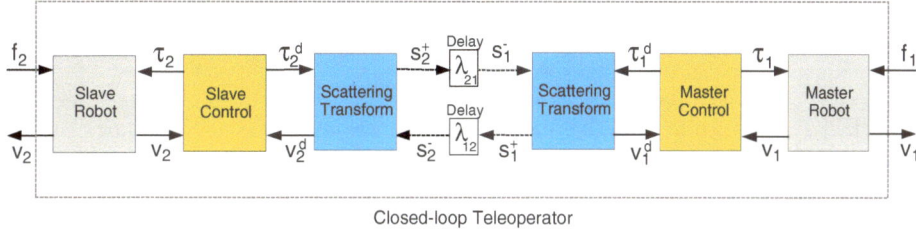

FIGURE 2. Scattering-based teleoperation with communication delay, where the scattering transform passifies the two-port delayed communication block.

domain, the scattering operator $S(s) \in \mathcal{C}^{2 \times 2}$ is defined by

$$\frac{(\tau_d - bv_d)}{\sqrt{2b}} = S(s)\frac{(\tau_d + bv_d)}{\sqrt{2b}}$$

where $s \in \mathcal{C}$ is the Laplace variable, $b > 0$ is the scaling factor (or characteristic wave impedance [15]), and the terms $s^- := (\tau_d - bv_d)/\sqrt{2b} =: [s_1^-, s_2^-]^T$ and $s^+ := (\tau_d + bv_d)/\sqrt{2b} =: [s_1^+, s_2^+]^T$ are the reflected and incident waves (adopting the notations of [14]). The scattering operator $S(s)$ is called BR [12], if it is stable and

$$I - S^*(jw)S(jw) \succeq 0 \quad \text{or, equivalently} \quad ||S||_\infty := \sup_w \sqrt{\lambda[S^*(jw)S(jw)]} \leq 1$$

for all $w \in \Re$, where $S^*(jw) := \overline{S(jw)}^T$ (i.e. Hermittian) and $A \succ B$ means $A - B$ is positive-definite.

The relation between the passivity and this BR of $S(s)$ can then be informally checked s.t.

$$\int_0^T [\tau_1^d v_1^d + \tau_2^d v_2^d]dt = \int_0^T \tau_d^T v_d dt = \frac{1}{2}\int_0^T \left(||s^+||^2 - ||s^-||^2\right) dt$$

$$= \frac{1}{4\pi}\int_{-\infty}^{+\infty} (\tau_d + v_d)^*[I - S^*(jw)S(jw)](\tau_d + v_d)dw \geq 0$$

where the Parseval's identity is used to go down to the second line, and the last inequality is obtained by using the BR condition of $S(s)$. If the master and the slave local control modules (after the scattering blocks) are also passive (e.g. PI (proportional-integral) control), this passivity condition for τ_d, v_d will then imply the controller passivity, and, consequently, the closed-loop passivity, since all the other components are all passive. This inequality, in fact, show the passivity of the combined communication delay blocks and scattering blocks.

Of course, we have not established the BR of the scattering matrix $S(s)$ yet. As shown in [11], this can be achieved by the following simple communication law of the scattering variables $(s_1^+, s_1^-, s_2^+, s_2^-)$ as opposed to that of the power conjugated variables $(\tau_1^d, v_1^d, \tau_2^d, v_2^d)$:

$$s_1^-(t) = s_2^+(t - \lambda_{21}), \quad s_2^-(t) = s_1^+(t - \lambda_{12})$$

where λ_{12} is the delay from the master to slave. Note also that, for the master scattering block, the inputs are (s_1^-, τ_1^d), while the outputs are (s_1^+, v_1^d). Similar also holds for the slave side. With this scattering communication, we can then show that:

$$S(s) = \begin{bmatrix} 0 & e^{-s\lambda_{21}} \\ e^{-s\lambda_{12}} & 0 \end{bmatrix}$$

with $||S||_\infty = 1$. Thus, with the above scattering communication, S becomes BR and follows the passivity of the closed-loop teleoperation system.

This scattering (or wave) framework has been virtually the only way to achieve passive teleoperation with constant communication delay. Yet, some of its shortcomings have hampered its full usage for the Internet haptic teleoperation. First, this scattering/wave framework, in its standard form, only allows constant delay and will in general lose passivity when the delay is varying (e.g. elongation/shrinkage of wave signals [27]). An extension was made in [28] for the case of varying delay. However, a large damping is required there to absorb non-passive energy generation by the varying delay.

The other problem of the scattering framework, particularly problematic with packet loss of the Internet communication, is its well-known position drift problem due to the lack of explicit position feedback. Interestingly, passivity of the scattering framework is robust against packet loss, if we use null signal for the lost scattering variable, since doing so only results in the decrease of the energy-in-fly within the communication channel [29]. Yet, the packet loss effect is still accumulated in the scattering integration and eventually appears as the master-position drift (and/or sudden position jump when the wave-integral is used due to the sudden jump from the null signal [30]). These two problems much restrict the usefulness of the scattering framework for the haptic Internet teleoperation.

Perhaps, the most well-known and simplest control law, that contains position feedback, is the P-control (i.e. spring) or **PD-control** (spring and damping), which can be written as:

$$\tau_1 = -B_e \dot{q}_1 - B(\dot{q}_1 - \dot{q}_2(t - \lambda_{21})) - K(q_1 - q_2(t - \lambda_{21}))$$

for the master system (similar also holds for the slave side). In fact, this PD control has been frequently used for delayed teleoperation from its incepts to nowadays, without much theoretical justification, and with the anecdote that, as long as the delay is small enough and K is not so large, it works fine or, at least, can maintain stability [31]. Recently, using the concept of controller passivity and Parseval's identity, it was shown in [10] that this surprisingly simple LTI (linear time-invariant) PD-control (6) can indeed enforce passivity with the constant delay, if the control gains are tuned s.t.

$$B_e \succcurlyeq \frac{\max(\lambda_{12} + \lambda_{21})}{2} K$$

even if the master and slave systems are nonlinear and as long as all the signals pertinent to the analysis are Fourier transformable.[2] This condition says that the local dissipation B_e needs to be large, if we want a large P-gain K (i.e. high performance) with a long round-trip delay $\lambda_{12} + \lambda_{21}$. On the other hand, the D-gain B in the PD-control can be omitted without affecting passivity. Some subsequent refinements [16, 17] ensue this PD-scheme of [10].

It is shown in [17] that varying-delay may also be incorporated into this PD-scheme while enforcing passivity. This PD scheme is also immune to the position drift problem of the scattering framework, since it contains explicit position feedback. However, unlike the scattering framework, how to prove passivity for this PD control in the presence of

---

[2]Lem. 2 of [10] is in fact to guarantee this Fourier transformability. If this transformability is assumed, which is likely to hold in practice, this Lem. 2 in [10] is not needed.

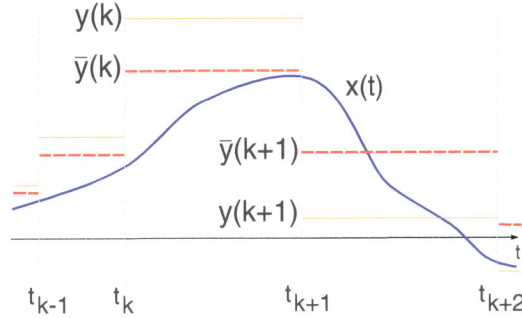

FIGURE 3. PSPM's hybrid setting: robot's continuous position $x(t)$, discrete set-position signal $y(k)$ received at t_k, and its modulated version $\bar{y}(k)$.

packet loss is evasive, although some experiments indicate that this may be the case, at least, under a certain condition [**32**]. A much more limiting aspect of this PD scheme for the Internet haptic teleoperation is, in fact, its performance. That is, if the round-trip delay $\lambda_{12} + \lambda_{21}$ gets longer (e.g. larger than 1 sec. [**10**]), the required damping injection B_e to attain a reasonable level of performance (i.e. K) will also need to be larger, resulting in a sluggish system. This is even more so for the case of pure varying-delay [**17**] or expected to be so for the Internet communication (i.e. by using the worst-case value among the randomly-varying round-trip delays). This large damping again much restricts the utility of the PD scheme for the Internet haptic teleoperation.

Thus, the common problem penetrating both the scattering and PD schemes for the haptic Internet teleoperation is the performance degradation due to the conservatism. We also want to avoid the position drift problem of the scattering framework and the breakdown of the passivity of the PD scheme with packet loss. The PSPM has been developed to address these issues: 1) by using P-action, it eliminates the position drift problem, as long as doing so does not violate passivity requirement; 2) by recognizing the hybrid nature (i.e. continuous + discrete time) and energetics associated to it, the PSPM enforces passivity with arbitrary packet loss and varying delay; and 3) by selectively modulating its control action (i.e. set-position) only when necessary according to the passivity requirement, it reduces the conservatism inherent to the constat gain state feedback scattering and PD control laws. We now provide more details on this PSPM. Our derivation of PSPM here is rather informal and intended to be more intuitive - refer to [**1, 3, 4**] for more detailed and rigorous development of PSPM.

4. Passive Set-Position Modulation

4.1. Hybrid Setting. Since the PSPM teleoperation framework to be presented here is symmetric w.r.t. the master and slave sites, and it is more efficient to derive the PSPM framework for a single system, let us consider the following single robotic system as the representative both for the master and slave systems:

$$(5) \qquad\qquad M(x)\ddot{x} + C(x,\dot{x})\dot{x} = \tau + f$$

where all the notations are the same as in (3)-(4). If the master and slave communicate their position signal via the Internet, its update rate would then be much slower than the local control servo-loop (typically faster than 1kHz).[3] Thus, we may consider such a desired set-position signal $y(k)$ as discrete-time signal, while the position $x(t)$ of the robot (5) continuous-time signal. This then creates the hybrid situation as depicted in Fig. 3, where the discrete set-position signal $y(k)$ is received (and jumps) at time t_k ($k = 1, 2, ...$), while $x(t)$ is continuous for all $t \geq 0$. Here, we do **not** assume that the intervals $I_k := [t_k, t_{k+1})$ are uniform nor that $y(k)$ was sent before $y(k + 1)$ from its sending port. This enables us to accommodate a variety of communication/data-update imperfectness of $y(k)$ including varying-delay, packet-loss, time-swapping, etc.

Then, as stated in the previous section, to have explicit position feedback, let us connect $x(t)$ and $y(k)$ via the following P-control with damping injection: for $t \in I_k = [t_k, t_{k+1})$,

$$(6) \qquad \tau(t) = -B\dot{x}(t) - K(x(t) - y(k))$$

where $B, K \in \Re^{n \times n}$ are the damping and P-control gains, each being symmetric and positive definite. It is well-known that the spring control in (6), directly connecting $x(t)$ and $y(k)$, can become unstable with varying-delay or packet-loss. As shown below, this is in fact due to the potentially **passivity-breaking energy jump** in the spring K with the switchings of $y(k)$.

More precisely, if we analyze the total energy of the closed-loop robotic system (5) under the control (6) from $I_0 = [t_0, t_1)$ with $t_0 = 0$ through $I_N = [t_N, t_{N+1})$, we can show that: for all $T \in [t_N, t_{N+1})$,

$$(7) \qquad V(T) - V(0) - \sum_{k=1}^{N} \Delta P(k) + \sum_{k=0}^{N-1} D(k) + \int_{t_N}^{T} ||\dot{x}||_B^2 dt = \int_0^T f^T \dot{x} dt$$

where $V(t) := \kappa(t) + \varphi(t)$ is the total energy with the kinetic energy $\kappa(t) := ||\dot{x}(t)||_M^2/2$ and the spring energy $\varphi(t) := ||x(t) - y(k)||_K^2/2$, with the notation $||x||_A := \sqrt{x^T A x}$ for a positive-definite and symmetric $A \in \Re^{n \times n}$;

$$(8) \qquad \Delta P(k) := \varphi(t_k) - \varphi(t_k^-) = \frac{1}{2}||x(t_k) - y(k)||_K^2 - \frac{1}{2}||x(t_k) - y(k-1)||_K^2$$

is the spring energy jump at t_k, which can be positive (i.e. passivity-breaking) or negative (i.e. passivity-enforcing); and

$$(9) \qquad D(k) := \int_{t_k}^{t_{k+1}} ||\dot{x}||_B^2 dt.$$

is the damping dissipation during $I_k = [t_k, t_{k+1})$.

Thus, if there is no energy jumping so that $\Delta P(k) = 0$, since $D(k) \geq 0$ and $V(t) \geq 0$, we can then have the closed-loop passivity (18) with $d^2 := V(0)$ by rearranging (7). However, if we have a certain sequence of $y(k)$, that creates positive net-energy jumping so that $\sum_{k=1}^{N} \Delta P(k)$ becomes excessively large, the closed-loop passivity will break down and the interaction stability/safety will be at risk. This clearly shows that the reason of passivity breaking for the P-control (6) is the spring energy jump due to the switchings of $y(k)$.

---

[3]This slow update assumption for $y(k)$ can be actually removed if the following condition is met:

$$B_e \geq B + \frac{B}{2}\left[1 + \frac{\Delta t_k}{\Delta t_{k-1}}\right] + \frac{K\Delta t_k}{2}$$

for all k, where B_e is the un-modeled device damping, and $\Delta t_k := t_{k+1} - t_k$. See [**22**] for more details.

4.2. PSPM Algorithm. The main idea of the PSPM is to modulate the raw set-position signal $y(k) \in \Re^n$ to its modulated version $\bar{y}(k) \in \Re^n$ in such a way that $\bar{y}(k)$ is as close to $y(k)$ as possible (i.e. performance) yet only to the extent permissible by the available energy in the system (i.e. passivity). We will also augment this basic modulation idea with some performance-enhancing mechanisms.

Let us first consider the interval $I_k = [t_k, t_{k+1})$. Then, we have (possibly passivity-breaking) $\Delta P(k)$ at t_k, yet, also (passivity-enforcing) $D(k)$ during I_k. This suggests us the following strategy: at t_k,

$$\min_{\bar{y}(k)} \ \|y(k) - \bar{y}(k)\|$$

$$\text{subj. } D(k) - \Delta \bar{P}(k) \geq 0$$

where

$$(10) \qquad\qquad \Delta \bar{P}(k) := \bar{\varphi}(t_k) - \bar{\varphi}(t_k^-)$$

is the spring energy jump at t_k when the modulation $\bar{y}(k)$ is used, with $\bar{\varphi}(t_k) := \frac{1}{2}\|x(t_k) - \bar{y}(k)\|_K^2$ and $\bar{\varphi}(t_k^-) := \frac{1}{2}\|x(t_k) - \bar{y}(k-1)\|_K^2$. Note that the second-line enforces passivity, by making the energy jump $\Delta \bar{P}(k)$ be all dissipated by $D(k)$.

Yet, this simple strategy is not implementable, since computing $D(k)$ in (9) at t_k requires future information, i.e. $\dot{x}(t)$ during $[t_k, t_{k+1})$. To avoid this causality problem, we "skew" the strategy s.t.: at t_k,

$$(11) \qquad\qquad \min_{\bar{y}(k)} \ \|y(k) - \bar{y}(k)\|$$

$$(12) \qquad\qquad \text{subj. } D(k-1) - \Delta \bar{P}(k) \geq 0$$

where $D(k-1)$ is now computable at t_k. Note that the new condition (second line) still enforces passivity, since the energy jump $\Delta \bar{P}(k)$ is now regulated less than (i.e. allowable by) the damping dissipation $D(k-1)$ in the previous $I_{k-1} = [t_{k-1}, t_k)$. We further refine this base-line algorithm:

(1) To avoid error-prone numerical differentiation/integration in computing $D(k)$, we use the following purely position-dependent estimation $D_{\min}(k)$: with diagonal B,

$$(13) \qquad D_{\min}(k) := \frac{1}{t_{k+1} - t_k} \sum_{i=1}^{n} b_i (x_i^{\max}(k) - x_i^{\min}(k))^2 \leq D(k)$$

where $b_i > 0$ is the i-th diagonal element of B, and $x_i^{\max}(k)$ and $x_i^{\min}(k)$ are the maximum and minimum of $x_i(t)$ during $I_k = [t_k, t_{k+1})$. For more detailed derivation of (13), see [**1, 4**]. As also shown in [**1**, Lem. 1], the estimation error $e_D(k) := D(k) - D_{\min}(k) \geq 0$ can be made practically negligible by simply densifying the data stream $y(k)$ (e.g. smoothing, interpolation, null data).

(2) To avoid "take-off" problem (i.e. the system does not move with zero initial energy), we augment the system with a virtual energy reservoir E. More precisely, we simulate (in computer) the following discrete-time dynamics of $E(k) \in \Re$: with $E(0) > 0$, at each t_k,

$$(14) \qquad\qquad E(k) = E(k-1) + D_{\min}(k-1) - \Delta \bar{P}(k) \geq 0$$

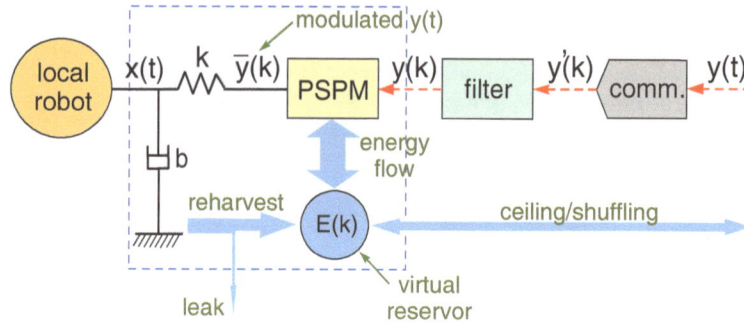

FIGURE 4. Energetics of PSPM, where $\bar{y}(k)$ is computed from $y(k)$ using re-harvested energy, received energy (from shuffling), and virtual reservoir energy. Note that any data processing can be inserted in "filter" block while enforcing passivity (e.g. estimator, predictor, or densifier (to reduce energy leak)).

implying that, at each t_k, $E(k)$ will provide energy to generate $\Delta \bar{P}(k)$ (or re-ceive energy from it), while also re-harvesting otherwise-wasted (major) portion of dissipation through B via $D_{\min}(k-1)$. This B dissipation is completely an artifact of the control (6) (i.e. not due to the real interaction with humans/environments). Thus, we want to recapture $D(k)$ as much as possible for better energy efficiency.

(3) We also implement energy shuffling/ceiling into PSPM: 1) along with $y(k)$, energy shuffling term $\Delta E_y(k) \geq 0$ is also transmitted and, accordingly, (14) is modified by

(15) $$E(k) = E(k-1) + \Delta E_y(k) + D_{\min}(k-1) - \Delta \bar{P}(k) \geq 0;$$

and 2) if $E(k) > \bar{E}$ after (15), we set $E(k) = \bar{E}$ (i.e. ceiling), and send $\Delta E_x(k) := E(k) - \bar{E}$ to its counter-part along with the set-position signal $x(t_k)$. This energy shuffling/ceiling, for instance, would then allow us to recharge the slave robot operating in highly-dissipative environments, by sending energy from the human operator, who, in teleoperation, usually keeps injecting energy into the system (to command the slave robot).

The complete PSPM algorithm is presented in Algo. 1. The optimization problem (16)-(17) always has a unique solution due to its convexity. In the line 3 of Algo. 1, we assume that the PSPM is triggered whenever a new data $(y, \Delta E_y)$ is received. Other triggering mechanisms may also be possible here (e.g. timed-trigger with $(y, \Delta E_y) = (y(k-1), 0)$). See also Fig. 4 for the energetics of this PSPM algorithm. In the following Th. 1, we present the main properties of this PSPM algorithm. Its proof we omit here. For its complete proof, see [4].

THEOREM 1. *Consider the robotic system* (5) *under the control* (6) *and the PSPM. Suppose that* $\dot{x}(t), x(t)$ *are continuous* $\forall t \geq 0$, *and the received energy shuffling terms are bounded, that is,* $\exists$ *a bounded constant* $b \in \Re$ *s.t.* $\sum_{j=1}^{k} \Delta E_y(j) \leq b^2 \ \forall k \geq 1$.

(1) *(Passivity/stability) The closed-loop system is passive in the sense of* (18). *Also, if the human/environment is passive, i.e.,* $\forall T \geq 0$, $\exists$ *a bounded constant* $c \in \Re$

Algorithm 1 Passive Set-Position Modulation

1: $\bar{y}(0) \Leftarrow x(0)$, $E(0) \Leftarrow \bar{E}$, $k \Leftarrow 0$
2: **repeat**
3:       **if** data $(y, \Delta E_y)$ received **then**
4:          $k \Leftarrow k + 1$
5:          $y(k) \Leftarrow y$, $\Delta E_y(k) \Leftarrow \Delta E_y$
6:          retrieve $x(t_k)$, $x_i^{\max}(k-1)$, $x_i^{\min}(k-1)$
7:          find $\bar{y}(k)$ by solving

(16) $\min_{\bar{y}(k)} \; \|y(k) - \bar{y}(k)\|$

(17)   subj. $E(k) \Leftarrow E(k-1) + \Delta E_y(k) + D_{\min}(k-1) - \Delta \bar{P}(k) \geq 0$

8:          **if** $E(k) > \bar{E}$ **then**
9:             $\Delta E_x(k) \Leftarrow E(k) - \bar{E}$, $E(k) \Leftarrow \bar{E}$
10:         **else**
11:            $\Delta E_x(k) \Leftarrow 0$
12:         **end if**
13:         send $(x(t_k), \Delta E_x(k))$ or discard if no counter-part
14:      **end if**
15: **until** operation terminated

---

     s.t.

(18) $$\int_0^T f^T \dot{x} dt \leq c^2$$

     interaction stability is guaranteed with bounded velocity $\dot{x}(t)$ $\forall t \geq 0$.

   (2) *(Position tracking) Assume $f(t) = 0$ and there is un-modeled damping B_e for the robot (5). Assume also that $\exists$ bounded $\lambda_1, \lambda_2 > 0$ s.t. $\underline{\sigma}[M(x)] \geq \lambda_1$ and $|\partial M_{qr}(x)/\partial x_s| \leq \lambda_2$ for all $x \in \Re^n$, $q, r, s = 1, ..., n$, where $\underline{\sigma}(\star)$ is the smallest singular value of $\star$, and M_{qr}, x_s are the qr^{-th} and s^{-th} components of M and x, respectively. Then, $\dot{x}(t) \rightarrow 0$, $\|\bar{y}(k+1) - \bar{y}(k)\| \rightarrow 0$, and $\|x(t) - \bar{y}(k)\| \rightarrow 0$. Suppose further that $\exists p \geq 0$ s.t. $E(k) > 0$ $\forall k \geq p$ and $\|y(k+1) - y(k)\| \rightarrow 0$. Then, $\|x(t) - y(k)\| \rightarrow 0$.*

   (3) *(Haptic feedback) Suppose that $(\ddot{x}(t), \dot{x}(t), x(t)) \rightarrow (0, 0, x_o)$ and $y(k) \rightarrow y_o$ with constant $x_o, y_o \in \Re^n$, and $\exists p \geq 0$ s.t. $E(k) > 0$ $\forall k \geq p$. Then, $f(t) \rightarrow K(x_o - y_o)$.*

     The key condition for the performance here is that $E(k)$ does not deplete energy. Of course, if the interacting environment/human behaves as a strongly dissipating system (e.g. huge damper), $E(k)$ will likely deplete and, eventually, performance will deteriorate. This is natural here, since it is how any passive system is supposed to work: restricts performance to enforce passivity. Even so, as shown in the below experiments, energy depletion of $E(k)$ does not occur often, and limiting performance by the PSPM algorithm happens fairly rarely (unless intentionally trying to induce this - see Fig. 7). This implies better performance of the PSPM as compared to other constant gain state feedback scattering and PD schemes, that will blindly restrict the performance, although the occasions requiring to do so are rare. See Sec. 4.3.

     Note that the dissipation $D(k)$ is through the (artificial) control damping B in (6), rather than through the real robot-environment/human interaction via the power-port $f^T \dot{x}$.

Thus, we want to minimize this internally-wasted energy dissipation. By doing so, we will be able to improve energy efficiency, and, consequently, attain better performance (i.e. more energy to push $\bar{y}(k)$ closer to $y(k)$). One way to do so is to improve the efficiency of the energy re-harvesting by reducing the "energy-leak" $\sum e_D(k) = \sum[D(k) - D_{\min}(k)]$. See Fig. 4. As shown in [**1**, Lem. 1], this energy leak is **quadratically** reduced as the update rate of $y(k)$ speeds up. This means that, for instance, if we ten-times densify a 50ms data-stream $y(k)$ by using a certain 5ms-running smoothing-filter (or even simply by putting null-data/duplicating previous data every 5ms), this energy leak can be reduced by 99% so that it would be practically negligible. Here, note that such a densifying process, when located in the "filter"-block in Fig. 4, will not jeopardize the passivity of the PSPM, since the PSPM only requires the incoming data-stream (to the "PSPM"-block in Fig. 4) to be discrete.

Now, we have enough machinery to apply the PSPM for the Internet haptic teleoperation. We simply implement the PSPM in the master and the slave sides in exactly the same manner (i.e. symmetric teleoperation architecture). Then, essentially, by combining the results of Th. 1 for the master and the slave sides, we can show the following Th. 2, which summarize properties/performance of the PSPM-based Internet haptic teleoperation. Here, $\star_1, \star_2$ (or $\star^1, \star^2$) represent variables for the master and slave sides, respectively. Again, the proof of this Th. 2 is omitted here and we refer readers to [**1,4**] for the complete proof.

THEOREM 2. *Consider master and slave robots* (3)-(4) *communicating over the Internet. Suppose we implement local spring-damper control* (6) *and the PSPM in Algo. 1 for each of them individually. Suppose further that there are no duplicated data receptions, and $x_i(t), \dot{x}_i(t)$ are continuous $\forall t \geq 0$, $i = 1, 2$.*

(1) *(Passivity/stability) The closed-loop teleoperator is two-port passive: $\forall T \geq 0$,*

$$\int_0^T [f_1^T \dot{x}_1 + f_2^T \dot{x}_2] dt \geq -d^2$$

where $d \in \Re$ is a bounded constant. Also, if the human operator and slave environment are individually passive (i.e. each satisfying (18)), the interaction is stable with bounded $\dot{x}_i(t)$ $\forall t \geq 0$, $i = 1, 2$.

(2) *(Position coordination) Assume $f_1 = f_2 = 0$ and un-modeled positive-definite dampings B_1^e, B_2^e for the master and slave robots. Assume also that $\exists$ bounded $\lambda_1^i, \lambda_2^i > 0$ s.t. $\underline{\sigma}[M_i(x_i)] \geq \lambda_1^i$ and $|\partial M_{qr}^i(x_i)/\partial x_i^s| \leq \lambda_2^i$ for all $x_i \in \Re^n$, $q, r, s = 1, ..., n$ and $i = 1, 2$, where M_{qr}^i, x_i^s are the qr^{-th} and s^{-th} components of M_i and x_i, respectively. Then, $\dot{x}_i \to 0$, $\|\bar{y}_i(k+1) - \bar{y}_i(k)\| \to 0$ and $\|x_i(t) \to \bar{y}_i(k)\| \to 0$, $i = 1, 2$. Moreover, if $\exists p \geq 0$ s.t. $E_i(k) > 0$ $\forall k \geq p$, $i = 1, 2$, $\|x_1(t) - x_2(t)\| \to 0$.*

(3) *(Force reflection) Suppose $(\ddot{x}_i(t), \dot{x}_i(t)) \to 0$, $(x_1(t), x_2(t)) \to (x_o^1, x_o^2)$, and $\exists p \geq 0$ s.t. $E_i(k) > 0$ $\forall k \geq p$, $i = 1, 2$, where $x_o^1, x_o^2 \in \Re^n$ are constant vectors. Then,*

(19) $$f_1(t) \to K_1(x_o^1 - x_o^2), \quad f_2(t) \to K_2(x_o^2 - x_o^1)$$

where K_1, K_2 are the spring gains for (6) for the master and slave, respectively.

At this point, it would be worthwhile to emphasize unique properties of the PSPM. First, due to its variable structure selectively modulate its control action (i.e. set-position), compared to the constant gain state feedback scattering and PD schemes, it can provide significantly better performance as shown in Sec. 4.3. Also, since this PSPM only requires the incoming data-stream $y(k)$ to be discrete, 1) it can enforce passivity even if the data-stream $y(k)$ embeds a variety of communication imperfectness in it (e.g. varying-delay, packet loss, and even time-swapping); and 2) it can accommodate *any* intermediate data-processing at the "filter"-block in Fig. 4, simply to densify the data-stream for better energy efficiency or to enhance performance (e.g. predictor/estimator), while still enforcing passivity. It is also local/decentralized in the sense that its implementation/tuning (e.g. choosing B, K) can be done without any consulting with communication counter-parts. This may be useful for some applications demanding scalability (e.g. multiuser haptic collaboration). Finally, in contrast to other teleoperation/haptics results (e.g. [10, 15, 23, 29]), the PSPM framework exploits the hybrid nature of the problem (i.e. continuous $x(t)$, discrete $y(k)$), rather than avoids/neglects it.

4.3. Teleoperation Experiment. We perform an experiment using Phantom Desktop as a master and Phantom Omni as a slave, both running with 1 ms low-level control servo-rate. For more details, also refer to [35]. A simple probabilistic model is used separately for forward (i.e. master to slave) and backward (i.e. slave to master) Internet communication. The following characteristics are achieved for the forward (or backward, resp.) communication: delay (from sending to receiving a data as measured by a common clock) randomly varies from 0.1s to 0.48s (or 0.11s to 0.56s, resp.); update interval I_k randomly varies from 1.2ms to 198.5ms (or 1.2ms to 223.6ms, resp.); and packet-loss rate is 93.31% (or 92.15%, resp.). For the PSPM, we use the following parameters: $(B_1, K_1, E_1(0), \bar{E}_1) = (5\text{Ns/m}, 100\text{N/m}, 0.015\text{Nm}, 0.03\text{Nm})$ for the master; and $(B_2, K_2, E_2(0), \bar{E}_2) = (5\text{Ns/m}, 50\text{N/m}, 0.015\text{Nm}, 0.03\text{Nm})$ for the slave. We also install a wall in the slave environment (around $x_2 = -4\text{mm}$).

We first perform the teleoperation task with no smoothing filter and the results are shown in Fig. 5, where it is clear that: 1) the interaction is stable between the human, the wall, and the closed-loop teleoperator over the Internet with varying-delay/packet-loss; 2) position coordination is achieved once the slave goes outside the wall; and 3) the human can perceive contact force during the hard-contact. Here, note that the (commanding) human operator keeps injecting energy into the system. This energy, then, is captured by the energy re-harvesting, quickly charges E_1 to $\bar{E}_1$, and, thenceforth, shuffles to the slave side (around 11.5s), resulting in sharp increase of E_2 around 12s (after E_2's initial plunge to drive the slave to follow the master). During the hard-contact, E_2 decreases. This energy from E_2 is mainly consumed to create a large deformation of the slave PSPM spring K_2 and the wall, and, when the contact is removed, flows back to E_2 (around 30s). Here, note that, even if E_2 depletes during a hard-contact, system performance (i.e. perception of wall at x_w) would not be much compromised. This is because, in this case, with $E_2 = 0$, x_2 will stuck with $x_2 \approx x_w$, thus, $\bar{y}_2(k+1) = \bar{y}_2(k) \approx x_w$. This implies that the human will still be able to perceive the wall at x_w with the same stiffness K_1, although f_1 may not be the same as f_2.

We also implement a 5ms-running smoothing low-pass-filter (situated at the "filter"-block in Fig. 4 for passivity), and achieve the results as shown in Fig. 6, where we can observe: 1) smoothing effect, particularly as shown by smoother master/slave forces (small sharp jumps in τ_2 are due to the stiction-like behavior due to device Coulomb friction, which becomes more dominant here as the operation slows down than in Fig. 5); 2) slower response due to the filter time-lag (e.g. longer operation time and higher τ_1 during free-motion due to slower change of $y_1(k)$); and 3) much better energy re-harvesting efficiency (e.g. decrease of E_2 during the contact almost fully compensated by re-harvested energy both from master and slave sides).

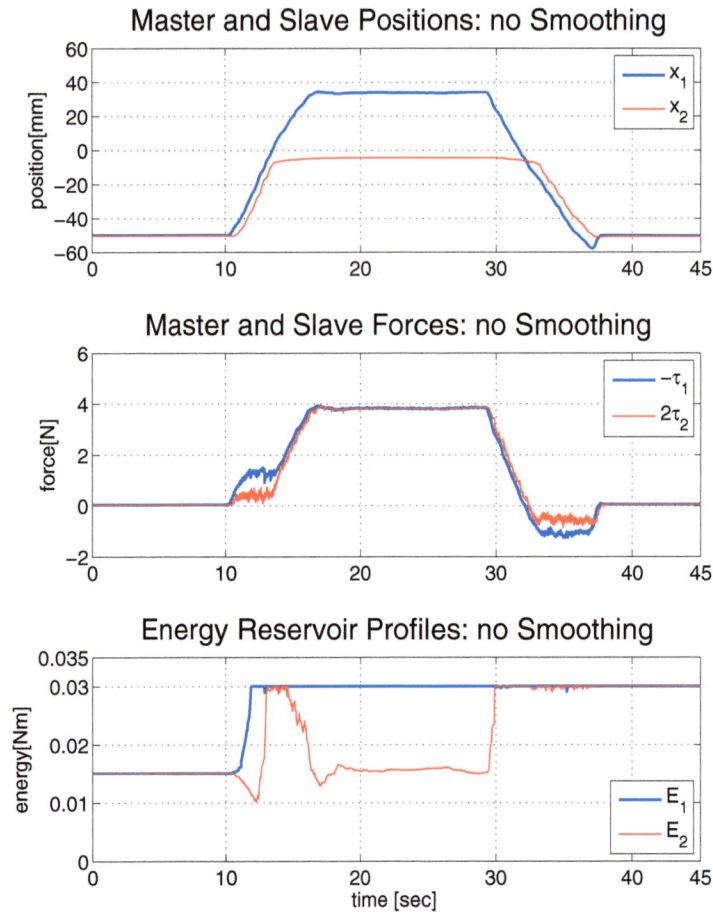

FIGURE 5. Teleoperation over the Internet. Notice: 1) interaction stability; 2) position coordination; 3) force reflection; and 4) human energy injection and energy shuffling.

During these experiments in Figs. 5 and 6, the energy reservoirs E_1, E_2 do not deplete during the operation with $y_\star(k) = \bar{y}_\star(k)$ for both the master and slave sides. This explains why the PSPM provides better performance (e.g. less damping) than the constant gain state feedback type scattering or PD frameworks (e.g. [10, 28]). That is, for those "static" scattering or PD frameworks, their performance-limiting mechanisms (e.g. large damping), designed for worst-case scenarios, are fully-activated all the time, while, for the PSPM, those mechanisms are activated only when necessary, or even not activated at all if not necessary as shown here in Figs. 5 and 6! Also, note that, the 5Ns/m damping of the PSPM used here is order-of-magnitude smaller than that required by the PD-scheme [10] (i.e. $B_1 \geq 0.5(0.48 + 0.56)K_1 \approx 50$Ns/m), thereby, providing much better performance. This is again possible here since, to enforce passivity, the PSPM selectively modulates its

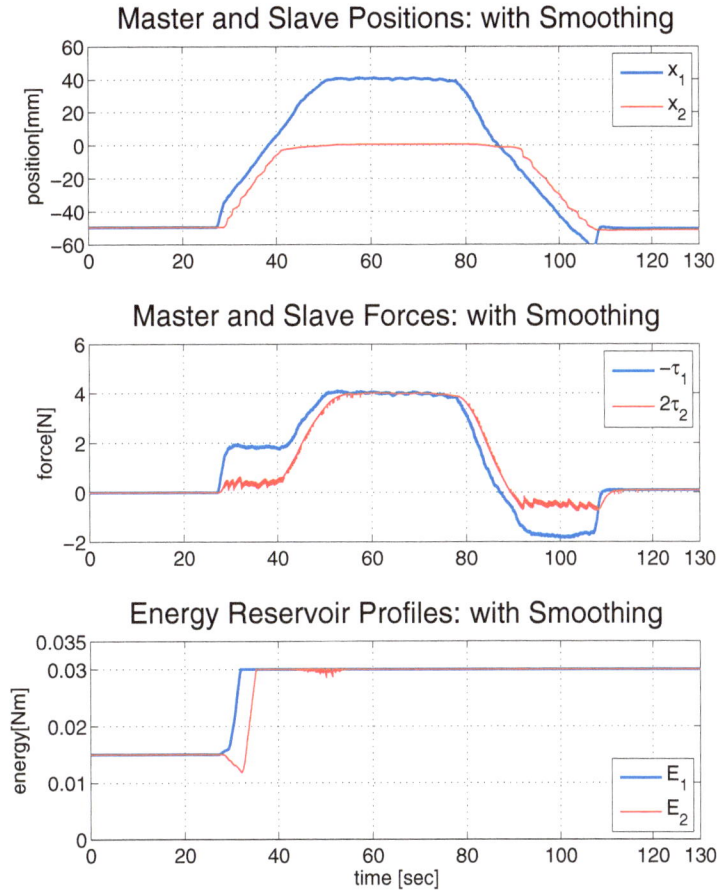

FIGURE 6. Teleoperation with smoothing/densifying filter. Notice: 1) smoother/slower system response; and 2) much better energy re-harvesting performance, as compared to Fig. 5.

action only necessary, rather than blindly applies large constant damping during the entire operation duration.

The last experiment is to show the (exacerbated) actions of the PSPM, that is, we intentionally tried to destabilize the system (yet, failed) - see Fig. 7, where $\times$ and $\bullet$ denote $y_i(k)$ and $\bar{y}_i(k)$ respectively. There, we can clearly see that, when the system starts to behave too aggressively (e.g. reaching passivity/stability margin), the PSPM kicks in (i.e. $E_i(k) = 0$ with less aggressive $\bar{y}_i(k) \neq y_i(k)$) to enforce passivity (thus, also stability).

5. Conclusion

In this book chapter, we introduce a recently proposed passive set-position modulation (PSPM) framework [1]- [4] as a means for the passive, yet, high-performance Internet haptic teleoperation. This PSPM selectively modulates its control action (i.e. set-position) only when necessarily, thereby drastically reducing the conservatism of the constant gain

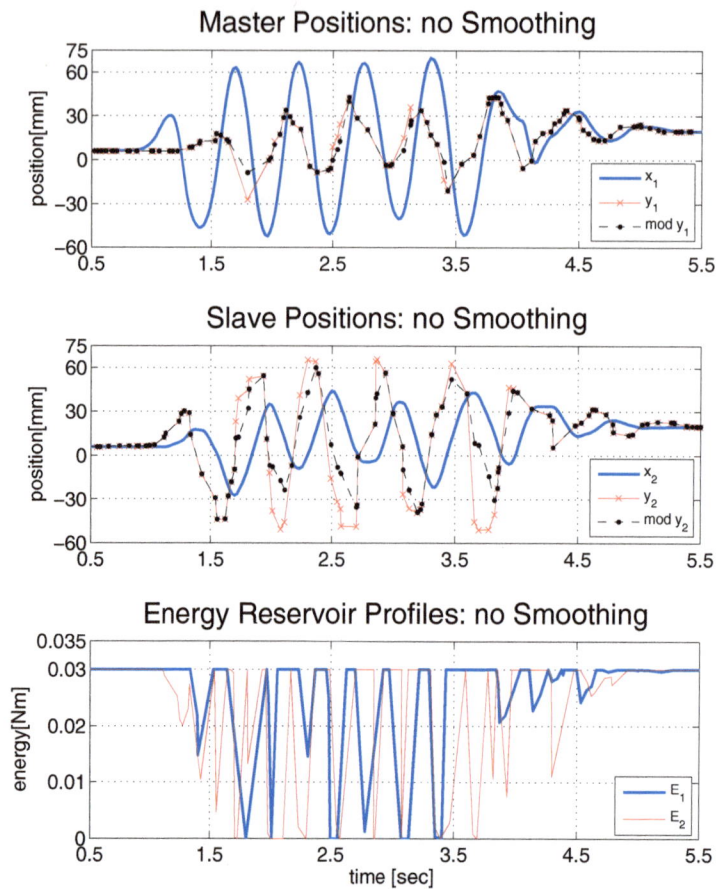

FIGURE 7. Passifying/stabilizing action of PSPM. Note that the PSPM produces less aggressive $\bar{y}(k)$ than $y(k)$, when the system starts to lose passivity (or stability).

state feedback type scattering and PD teleoperation schemes and providing much improved performance. Some contextuation of this PSPM w.r.t. the state of the art in haptics and delay teleoperation are also provided. Internet haptic teleoperation experimental results with this PSPM implemented are also presented.

We believe that this PSPM is promising for many applications, where passive high-performance teleoperation is needed/instrumental. For this, the following properties of the PSPM seem particularly powerful: 1) intermediate data processing (at the "filter" block in Fig. 4) to enhance the system performance (e.g. predictive display [33], estimator) while enforcing passivity, thereby, possibly challenging the presumption that passivity implies poor performance; and 2) locality/decentralizability of the PSPM for applications where scalability is demanded and the communication link is not always bilateral (e.g. large-scale multiuser haptic collaboration over the Internet [34]). Some works have been in progress in our laboratory toward these directions.

Acknowledgement

Research supported in part by the National Science Foundation under the Grant CMMI-0727480.

Bibliography

[1] D. J. Lee and K. Huang. Passive position feedback over packet-switching communication network with varying-delay and packet-loss. In *Symposium of Haptic Interfaces for Virtual Environments & Teleoperator Systems*, pages 335–342, 2008.

[2] D. J. Lee. Semi-autonomous teleoperation of multiple wheeled mobile robots over the internet. In *ASME Dynamic Systems & Control Conference*, 2008. Available at http://web.utk.edu/~djlee/papers/DSCC08a.pdf.

[3] D. J. Lee and K. Huang. Passive set-position modulation approach for haptics with slow variable, and asynchronous update. In *World Haptics Conference*, 2009. Available at http://web.utk.edu/~djlee/papers/Haptics09.pdf.

[4] D. J. Lee and K. Huang. Passive set-position modulation framework for interactive robotic systems. *IEEE Transactions on Robotics*. Submitted. Manuscript available at http://web.utk.edu/~djlee/papers/LeeKeTRO09.pdf.

[5] D. J. Lee, O. Martinez-Palafox, and M. W. Spong. Bilateral teleoperation of a wheeled mobile robot over delayed communication networks. In *Proceedings of IEEE International Conf. on Robotics & Automation*, pages 3298–3303, 2006.

[6] J. Kim, H. Kim, B. K. Tay, M. Muniyandi, M. A. Srinivasan, J. Jordan, J. Mortensen, M. Oliveira, and M. Slater. Transatlantic touch: a study of haptic collaboration over long distance. *Presence*, 13(3):328–337, 2004.

[7] G. Sankaranarayanan and B. Hannaford. Virtual coupling schemes for position coherency in networked haptic environments. In *Proceedings of the IEEE Int'l Conf. on Biomedical Robotics & Biomechatronics*, 2006.

[8] M. W. Spong and N. Chopra. Synchronization of networked lagrangian systems. In *Lecture Notes in Control &Information Sciences (vol 366, Proc. of 2006 Lagrangian & Hamiltonian Methods for Nonlinear Control)*, 2007.

[9] S. P. Buerger and N. Hogan. Complementary stability and loop shaping for improved human-robot interaction. *IEEE Transactions on Robotics*, 23(2):232–244, 2007.

[10] D. J. Lee and M. W. Spong. Passive bilateral teleoperation with constant time delay. *IEEE Transactions on Robotics*, 22(2):269–281, 2006.

[11] R. J. Anderson and M. W. Spong. Bilateral control of tele-operators with time delay. *IEEE Transactions on Automatic Control*, 34(5):494–501, 1989.

[12] B. Anderson and S. Vongpanitlerd. *Network analysis and synthesis*. Prentice Hall, Englewood Cliffs, NJ, 1973.

[13] G. Niemeyer and J. J. E. Slotine. Stable adaptive teleoperation. *IEEE Journal of Oceanic Engineering*, 16(1):152–162, 1991.

[14] S. Stramigioli, A. van der Schaft, B. Maschke, and C. Melchiorri. Geometric scattering in robotic telemanipulation. *IEEE Transactions on Robotics and Automation*, 18(4):588–596, 2002.

[15] G. Niemeyer and J. J. E. Slotine. Telemanipulation with time delays. *International Journal of Robotics Research*, 23(9):873–890, 2004.

[16] E. Nuno, R. Ortega, N. E. Barabanov, and L. Basanez. A globally stable pd-controller for bilateral teleoperators. *IEEE Transactions on Robotics*, 24(3):753–758, 2008.

[17] H. Kawada and T. Namerikawa. Bilateral control of nonlinear teleoperation with time varying communication delays. In *Proc. of the American Control Conference*, pages 189–194, 2004.

[18] J. E. Colgate and J. M. Brown. Factors affecting the z-width of a haptic display. In *Proceedings of the 1995 IEEE International Conf. on Robotics & Automation*, pages 3205–3210, 1994.

[19] J. E. Colgate, M.C. Stanley, and J. M. Brown. Issues in the haptic display of tool use. In *Proceedings of IEEE/RSJ International Conf. on Intelligent Robots and Systems*, volume 3, pages 140–145, 1995.

[20] R. B. Gillespie and M. R. Cutkosky. Stable user-specific haptic rendering of the virtual wall. In *Proceedings of ASME International Mechanical Engineering Congress and Exposition*, pages 397–406, 1996.

[21] J. E. Colgate and G. Schenkel. Passivity of a class of sampled-data systems: application to haptic interfaces. *Journal of Robotic Systems*, 14(1):37–47, 1997.

[22] D. J. Lee. Extension of colgate's passivity condition to variable-rate haptics. In *Proc. of IEEE/RSJ Int'l Conf. on Intelligent Robots & Systems*, pages 1761–1766, 2009.

[23] B. Hannaford and J-H. Ryu. Time domain passivity control of haptic interfaces. *IEEE Transactions on Robotics and Automation*, 18(1):1–10, 2002.

[24] D. J. Lee and P. Y. Li. Passive bilateral control and tool dynamics rendering for nonlinear mechanical teleoperators. *IEEE Transactions on Robotics*, 21(5):936–951, 2005.

[25] J-H. Ryu, B. Hannaford, D-S. Kwon, and J-H. Kim. A simulation/experimental study of the noisy behavior of the time-domain passivity controller. *IEEE Transactions on Robotics*, 21(4):733–741, 2005.

[26] J-H. Ryu, Y. S. Kim, and B. Hannaford. Sampled- and continuous-time passivity and stability of virtual environment. *IEEE Transactions on Robotics*, 20(4):772–776, 2005.

[27] K. Kosuge, H. Murayama, and K. Takeo. Bilateral feedback control of telemanipulators via computer network. In *Proceedings of IEEE/RSJ International Conf. on Intelligent Robots and Systems*, pages 1380–1385, 1996.

[28] N. Chopra, M. W. Spong, S. Hirche, and M. Buss. Bilateral teleoperation over the internet: the time varying delay problem. In *Proceedings of American Control Conference*, pages 155–160, 2003.

[29] P. Berestesky, N. Chopra, and M. W. Spong. Discrete time passivity in bilateral teleoperation over the internet. In *Proceedings of IEEE International Conf. on Robotics & Automation*, pages 4557–4564, 2004.

[30] G. Niemeyer and J. J. E. Slotine. Towards force-reflecting teleoperation over the internet. In *Proceedings of IEEE International Conf. on Robotics & Automation*, pages 1909–1915, 1998.

[31] R. Oboe and P. Fiorini. A design and control environment for internet-based telerobotics. *International Journal of Robotics Research*, 17(4):433–449, 1998.

[32] E. J. Rodriguez-Seda, D. J. Lee, and M. W. Spong. An experimental comparison of bilateral internet-based teleoperation. In *Proceedings of CCA/CACSD/ISIC*, 2006.

[33] W. S. Kim and A. K. Bejczy. Demonstration of a high-fidelity predictive/preview display technique for telerobotic servicing in space. *IEEE Transactions on Robotics and Automation*, 9(5):698–702, 1993.

[34] G. Sankaranarayanan and B. Hannaford. Experimental internet haptic collaboration using virtual coupling schemes. In *Proceedings of the IEEE Symposium on Haptic Interfaces for Virtual Environments & Teleoperator Systems*, 2008.

[35] K. Huang and D. J. Lee. Implementation and experiments of passive set-position modulation framework for interactive robotic systems. In *Proc. of IEEE/RSJ Int'l Conf. on Intelligent Robots & Systems*, pages 5615–5620, 2009.

Send Orders of Reprints at reprints@benthamscience.net

CHAPTER 11

STABLE TELEOPERATION WITH TIME DOMAIN PASSIVITY CONTROL[1]

Jee-Hwan Ryu

Korea University of Technology and Education

Cheonan, Chungnam, Korea

Dong-Soo Kwon

Korea Advanced Institute of Science and Technology

Daejeon, Korea

and

Blake Hannaford

University of Washington

Seattle, WA 98195, USA

[1]This article has been published in the IEEE Transactions on Robotics and Automation, Vol. 20, No. 2, pp. 365-373, 2004.

ABSTRACT. A bilateral control scheme is introduced to ensure stable teleoperation under a wide variety of environments and operating speeds. System stability is analyzed in terms of the time domain definition of passivity. A previously proposed energy-based method is extended to a 2-port network, and the issues in implementing the "Passivity Observer" and "Passivity Controller" to teleoperation systems are studied. The method is tested with our two-DOF master/slave teleoperation system. Stable teleoperation is achieved under conditions such as hard wall contact (stiffness > 150 kN/m) and hard surface following.

1. Introduction

The goal of teleoperation system control is to achieve transparency while maintaining stability (i.e., such that the system does not exhibit vibration or divergent behavior) under any operating conditions and for any environments. To this end, several bilateral control architectures have thus far been developed [7], [11], [13], [22], [23], and [27].

In designing the bilateral controller, a classic engineering trade-off between transparency and stability has been an important issue, since transparency must often be reduced in order to guarantee stable operation in the wide range of environment impedances (for example, in terms of stiffness of "free space" and "hard contact"). This has necessitated investigating into methods to increase transparency without introducing instability. Several previous studies have sought out theoretical design methods for control parameters based on linear circuit theory [1], [12] or linear robust control theory [4], [18], [26].

However, the teleoperation systems of our interest are non-linear and the dynamic properties of a human operator are always involved. These factors make it difficult to analyze teleoperation systems in terms of known parameters and linear control theory. To cope with the non-linearity and uncertain parameters of the teleoperation system, several researchers have used non-linear control laws such as adaptive control to design the bilateral controller [10], [17], [21], [29]. However, this approach requires, at the very least, system dynamic equations, and the system uncertainty should be captured with a few unknown parameters. Generally, it is very difficult to obtain an exact dynamic model of the teleoperation system. Furthermore, the dynamic structure of a teleoperation system is too complicated to capture with just a few parameters. Thus it becomes very complicated to apply this model-based approach when the teleoperation system has high Degrees of Freedom (DOF).

One promising approach is the use of the idea of passivity to guarantee stable operation without exact knowledge of model information. Anderson and Spong [3] and Neimeyer and Slotine [20] have used passivity concepts for stable teleoperation when a time delay exists. Yokokohji et al. [28] have introduced an energy monitoring method to satisfy passivity under time-varying communication delay. Lozano et al. [19] also presented an idea to solve the time-varying delay problem based on passivity. Lee and Li [15], [16] proposed a method to make the teleoperation system passive using fictitious energy storage. Colgate and Schenkel [5] have used passivity to derive fixed parameter virtual coupling (i.e., haptic interface controllers). Anderson [2] has implemented passive module idea to teleoperation systems. However, the use of the passivity for designing teleoperation systems has resulted in an overly conservative controller since they have analyzed system passivity in frequency domain which could not avoid fixed damping design. Thus, in many cases, performance can be poor since a fixed damping value is derived for guaranteeing passivity under all operating conditions. Recently, Hannaford and Ryu [8] have proposed a new energy based

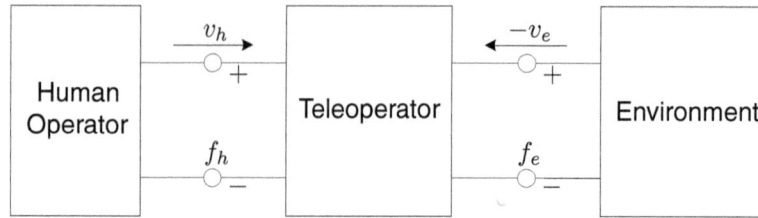

FIGURE 1. 2-port network representation of teleoperation systems. System blocks are (left to right) human operator, teleoperator, and environment.

method for stable haptic interaction, and proved and experimentally tested stability with minimal performance loss for 1-port haptic interaction.

This Chapter introduces a method to implement the time domain passivity control approach to teleoperation systems. The design method is extended to the 2-port network, and implementation issues are investigated.

2. Network model and stability condition

Fig. 1 shows a network model of a teleoperation system, where v_h and v_e denote the velocities at the interacting points of the human/master and environment/slave, respectively, and f_h and f_e represents the force that the operator applies to the master manipulator and the slave manipulator applies to the environment, respectively.

To investigate the stability of the teleoperation system, we analyze the network model based on the idea of passivity. If individual blocks of the network model are passive, the overall system is passive, and it is sufficient to make the system stable [6]. If we assume that the operator and the environment are passive, the teleoperator 2-port must be passive to meet the sufficient condition for stability [3], [27]. Generally, the environments are passive, and based on Hogan's experiment [9], we can assume that the human operator can be modeled as a passive network. Thus, we only need to make the teleoperator 2-port passive to satisfy the stability of the teleoperation system.

We then use the following widely known definitions of passivity for a multi-port network.

Definition. An M-port network with initial energy storage $E(0)$, is passive, if and only if

$$(1) \qquad \int_0^t \Big(f_1(\tau)v_1(\tau) + \cdots + f_M(\tau)v_M(\tau) \Big) \, d\tau + E(0) \geq 0, \forall t \geq 0$$

for admissible forces $f_1, \ldots, f_M$ and $v_1, \ldots, v_M$ velocities. The sign convention for all forces and velocities is defined to make their product positive when power enters the system port (Fig. 2b). The system is also assumed to have initial stored energy at $t = 0$ of $E(0)$. Equation 1 states that the energy supplied to a passive network must be greater than negative $E(0)$ for all time [24], [25].

3. Review of the Time-domain Passivity Control

This section reviews the "Passivity Observer" and "Passivity Controller", which have been applied to a 1-port network for application to haptic interfaces [8]. For a 1-port network (Fig. 2a) with zero initial energy storage, if we can measure the conjugate variables (f and v), which define power flow into a system, and the sampling rate is substantially faster

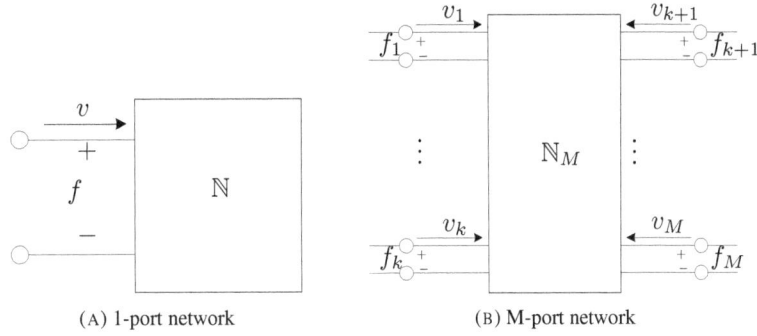

(A) 1-port network   (B) M-port network

FIGURE 2. 1-port and M-port networks representing components.

than the dynamics of the system so that the change in force and velocity with each sample is small, we can easily instrument one or more blocks in the system with the following "Passivity Observer," (PO) to measure energy flow into the 1-port network in real-time,

$$(2) \qquad E_{obsv}(n) = \Delta T \sum_{k=0}^{n} f(k)v(k)$$

where δT is the sampling period. If $E_{obsv}(n) \geq 0$ for every n , this means the system store and/or dissipates energy. If there is an instance where $E_{obsv}(n) < 0$, this means the system generates energy and the amount of generated energy is $-E_{obsv}(n)$. Note that it is generally possible to consider that the initial energy of a network system is zero since we start the operation with no initial deflection and errors for haptic and teleoperation systems.

Consider a 1-port system which may be active. Depending on operating conditions and the specifics of the 1-port elementâĂŹs dynamics, the PO may or may not be negative at a particular time. However, if it is negative at any time, we know that the 1-port may then be contributing to instability. Moreover, we know the amount of generated energy and we can design a time-varying element to dissipate only the required amount of energy. We call this element a "Passivity Controller" (PC) [**8**].

4. Time-domain Passivity Control for 2-port Networks

This section extends the PO and PC to a 2-port network for guaranteeing the stability of a teleoperation system by making the teleoperator 2-port passive. Similar to the 1-port case, the PO can be designed for a 2-port network (Fig. 5)

$$(3) \qquad E_{obsv}(n) = \Delta T \sum_{k=0}^{n} \Big(f_1(k)v_1(k) + f_2(k)v_2(k) \Big) = \Delta T \cdot W(n)$$

For designing PC, unlike 1-port network, two points should be considered.

(1) Add the PC at each port

(2) Activate the PC at active port

If only one PC is placed at either port (letâĂŹs assume first port), there might be some instance $E_{obsv}(n) < 0$, even though $v_1 = 0$ (or $f_1 = 0$). In that case, the generated energy at the other port cannot be dissipated since the only one PC at the first port cannot

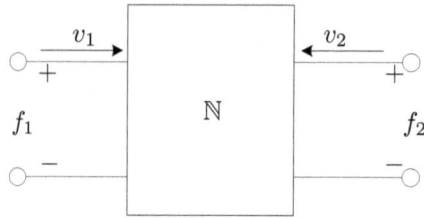

FIGURE 3. 2-port networks representing components.

TABLE 1. Several cases of two series PCs

Case	1	2	3	4.1	4.2
$\alpha_1(n)$	0	$\frac{-W(n)}{v_2^2(n)}$	0	$\frac{-f_2(n)v_2(n)}{v_2^2(n)}$	$\frac{-W(n)}{v_2^2(n)}$
$\alpha_2(n)$	0	0	$\frac{-W(n)}{v_3^2(n)}$	$\frac{-(W(n-1)+f_3(n)v_3(n))}{v_3^2(n)}$	0

be activated with the zero input signal. Consequently, another PC should be placed at the other port to dissipate the active energy output.

In addition, we have to consider how to activate the PC at each port to make the 2-port network passive. Mathematically, there are two ways to make the 2-port network passive (the total sum of energy is greater than zero).

- Make the produced energy less than the absorbed energy: activate the PC at the active port to cut the active energy output.
- Make the absorbed energy greater than the produced energy: activating the PC at the passive port to absorb more energy to satisfy this mathematical condition.

However, increasing the absorbed energy to make the network passive is not reasonable in a physical sense. For example, if the PC is placed between the 2-port network and an infinite energy source (such as effort or flow source), and the PC increases the absorbed energy to make the 2-port network passive, the total system energy may not be bounded. Thus, it is more feasible to make the produced energy less than the absorbed energy by activating the PC only at the active port. Note that it is possible to know which port is active or passive by monitoring the conjugate signal pair of each port in real time such as

$$(4) \qquad \textit{status of a port} = \begin{cases} \text{Active} & \text{if } f \cdot v < 0 \\ \text{Passive} & \text{if } f \cdot v \geq 0 \end{cases}$$

Based on the above, for a 2-port network with impedence causality at each port, we can design two series PCs (Fig. 4) in real time as follows:

(1) $v_1(n) = v_2(n)$ and $v_3(n) = v_4(n)$
(2) $f_2(n)$ and $f_3(n)$ are the outputs of the system
(3) $W(n) = \sum_{k=0}^{n-1} \left(f_1(k)v_1(k) + f_4(k)v_4(k) \right) + f_2(k)v_2(k) + f_3(k)v_3(k)$ is the PO
(4) Two series PCs can be designed for several cases in Table 1
(5) $f_1(n) = f_2(n) + \alpha_1(n)v_2(n) \Longrightarrow$ output
(6) $f_4(n) = f_3(n) + \alpha_2(n)v_3(n) \Longrightarrow$ output

where each case is as follows:

Case 1 $W(n) \geq 0$

In this case, energy does not flow out. There is no need to activate any PC.

Case 2 $W(n) < 0, f_2(n)v_2(n) < 0, f_3(n)v_3(n) \geq 0$

In this case, energy flows out from the left port. We need to activate only the left PC.

Case 3 $W(n) < 0, f_2(n)v_2(n) \geq 0, f_3(n)v_3(n) < 0$

In this case, energy flows out from the right port. We need to activate only the right PC.

Case 4 $W(n) < 0, f_2(n)v_2(n) < 0, f_3(n)v_3(n) < 0$

In this case, energy flows out from the both ports. To correct this active behavior, the PC must supply dissipation equal to $= W(n)$. For a impedance causality, we must allocate this damping among the 2-ports such that $\alpha_1(n)v_1^2(n) + \alpha_2 v_2^2(n) = -W$ We will give an example in which we prefer to allocate damping to the left port and the remainder to the right, but many other strategies are possible. The first case is when the produced energy from the right port is greater than the previously dissipated energy:

Case 4.1 $W(n-1) + f_3(n)v_3(n) < 0$

In this case, we only have to dissipate the net generation energy of the right port as the fifth column in Step 4. The second case is when the produced energy from the right port is less than the previously dissipated energy.

Case 4.2 $W(n-1) + f_3(n)v_3(n) \geq 0$

In this case, we do not need to activate the right port PC, and also reduce the conservatism of the left port PC as the sixth column in Step 4.

We can demonstrate that the system computed by Step 4 is passive in the following:

(5)

$$\sum_{k=0}^{n} (f_1(k)v_1(k) + f_4(k)v_4(k))$$

$$= \sum_{k=0}^{n} f_2(k)v_2(k) + \sum_{k=0}^{n} f_3(k)v_3(k) + \sum_{k=0}^{n} \alpha_1(k)v_2^2(k) + \sum_{k=0}^{n} \alpha_2(k)v_3^3(k)$$

$$= \sum_{k=0}^{n} f_2(k)v_2(k) + \sum_{k=0}^{n} f_3(k)v_3(k) + \sum_{k=0}^{n-1} \alpha_1(k)v_2^2(k) + \sum_{k=0}^{n-1} \alpha_2(k)v_3^3(k) + \alpha_1(k)v_2^2(k) + \alpha_2(k)v_3^3(k)$$

$$= W(n) + \alpha_1(k)v_2^2(k) + \alpha_2(k)v_3^3(k)$$

Using Step 4, we can make the 2-port network passive under all conditions,

$$\sum_{k=0}^{n} \left(f_1(k)v_1(k) + f_4(k)v_4(k) \right) \geq 0 \quad \forall n$$

The case of admittance causality can be similarly derived with a parallel PC. When the PC design is extended for a multi-port network, the cases are increased by 2^N, where N is the number of ports. However, the basic design rule is not changed. The PC is attached at each port of the network, and the PC is activated when the network become active and energy is produced at this port. One thing to determine is how to distribute the damping considering the previous energy dissipation when energy flows out through more than one port. One of the possibilities is distributing the damping proportional to velocity (or force)

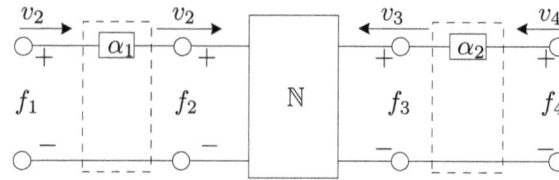

FIGURE 4. Series configurations of passivity controller for 2-port networks. α_1 and α_2 are adjustable damping elements at each port. Choice of configuration depends on input/output causality of model underlying each port.

at the port. We are currently studying the optimal damping distribution. Note that the damping at the active port at most reduces forces to zero (not negative).

5. Implementation Issues of Teleoperation Systems

This section addresses how to implement the developed 2-port PO/PC to teleoperation systems. There are two issues to be considered for implementing PO/PC.

5.1. Location of the PO/PC. First, to determine the place to locate the PO/PC, it is necessary to check the real-time availability of the conjugate signal pair at each port. In addition to the real-time availability, the conjugate output (which depends on causality) should be changed to a desired value in real-time for implementing the PC. Since our goal is making the teleoperator 2-port passive, it seems reasonable to place the PO/PC at teleoperator 2-port. If the forces f_h, f_e and the velocities v_h, v_e can be measured in real-time at both ports, it is usually possible to construct the PO (Eq. 3). However, these signals cannot be modified in real-time since these are responses of a physical interaction between human/environment and the teleoperator.

In this case, we can exclude some passive blocks until constructing an accessible conjugate pair without ruining the overall passivity. Physically, energy is transmitted to a physical system through the place where an actuator is placed, and the physical energy which is transmitted through the actuator can be calculated with the conjugate pair which define power flow from the actuator, such as force output from the actuator and the velocity at the actuating position. Based on this fact, we can divide the teleoperator into three parts; the mechanical part of the master manipulator, master/slave controller and communication channel, and the mechanical part of the slave manipulator. Fig. **??** shows a complete network model of a teleoperation system which is representing actual physical energy flows. The bilateral controller exchanges energy with the mechanical parts of the master/slave manipulator, and this energy flow can be measured with the conjugate pairs (f_m, v_m and f_s, v_s). f_m and f_s are actuator driving forces of the mechanical part of the master and slave manipulators, and v_m and v_s are velocities at the actuating place. The mechanical parts of the master and slave manipulator can be excluded since these do not make the teleoperator active due to their inherent dissipative elements. Thus it is sufficient to make the bilateral controller passive for making the teleoperator passive. To make the bilateral controller passive, the PO is designed as

$$E_{obsv}(n) = \Delta T \sum_{k=0}^{n} \Big(f_m(k)v_m(k) + f_s(k)v_s(k) \Big) = \Delta T \cdot W(n)$$

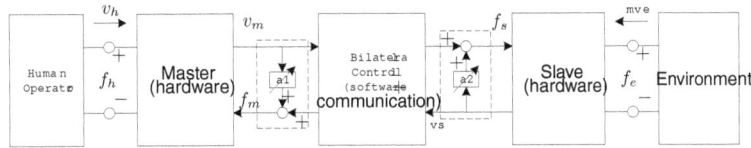

FIGURE 5. Block diagram of a complete teleoperation system. System blocks are (left to right) human operator, hardware part of master, bilateral controller including control software and communication channel, hard ware of slave, and environment.

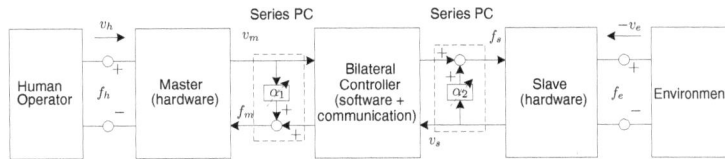

FIGURE 6. Block diagram of a complete teleoperation system with Passivity Controller. Two series PCs are attached at each port of bilateral controller.

and placed at the bilateral controller 2-port. As we intended, the conjugate pairs (f_m, v_m and f_s, v_s) can be accessible in real-time. Note that the PC is not included in the current PO iteration (n) since the PC is calculated based on the current PO value (refer step 3 of the PC algorithm).

5.2. Type of the PC. To determine the type of the passivity controller, it is required to determine the causality of each port of the bilateral controller. Usually motors are used for the actuator of the master and the slave manipulators, and motors have admittance causality, the bilateral controller has impedance causality at both ports. The inputs of the bilateral controller could be the velocity of the master v_m and the slave v_s, and the outputs of the bilateral controller are control inputs of the master f_m and the slave f_s at each port. Thus, the series type is the suitable configuration of the PC to absorb the energy output from the bilateral controller. Fig. 6 shows the teleoperation system with the PC. Note that the series PC appears to be connected in parallel, but this is an artifact of switching to block diagram notation for the connections between the master/slave, PC, and the bilateral controller.

6. Experimental Results

Based on the above studies, the PO and PC is implemented in a teleoperation system that has a two-DOF master (Fig. 7a) and a two-DOF slave manipulator (Fig. 7b). Master and slave manipulator have five-bar mechanism, and can display $2\ Nm$ for each joint axis. Each joint axis of the master and slave senses position in $1.6716 \cdot 10^{-4}\ rad$ and $1.6519 \cdot 10^{-4}\ rad$ increments, respectively. As a high stiffness environment, a steel wall is placed parallel to the Y-Axis. This system is entirely synchronous at $1000\ Hz$.

To experimentally study the PO/PC, we used a traditional bilateral control system (Position-Force) in which a position command is sent from master to slave and a force command is returned from slave to master. The slave robot has a PD control law for end effector position. In such an architecture, transparency is reduced by the damping supplied

(A) Master manipulator

(B) Slave manipulator

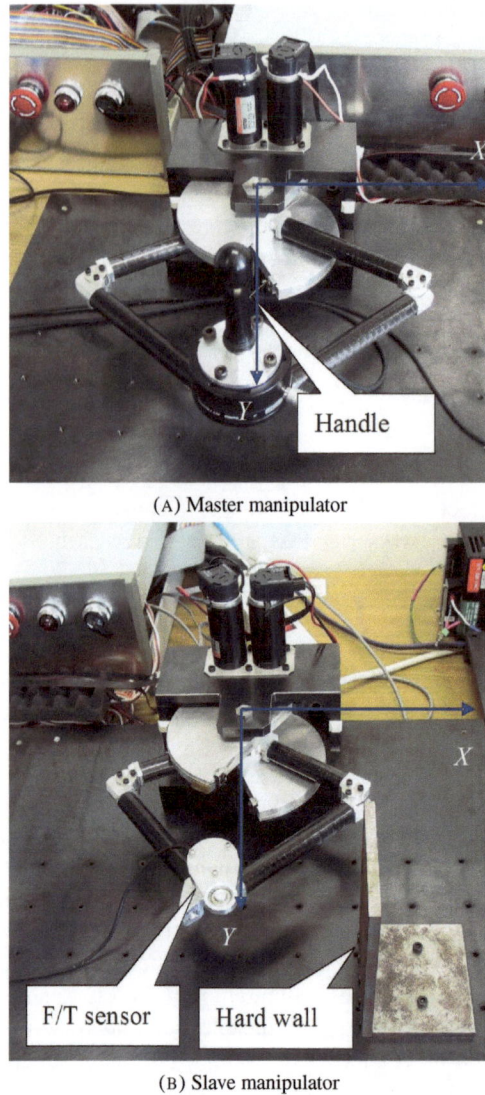

FIGURE 7. Two-DOF master manipulator (a), and two-dof slave manipulator (b) for teleoperation.

by the D term in the slave controller. For these experiments we increased transparency
beyond the normal limit - causing frequent unstable operation. Then the PC was added to
provide transient stabilization when needed.

6.1. Contact with High Stiffness. In the first experiment, without the PC, the operator maneuvered the master to make the slave contact with the hard wall (approximately
over 150 kN/m) with a relatively high velocity of about 120 mm/s. This resulted in an
oscillation observable as force and position pulses (Fig. 8a, b). The value of the PO was
initially positive, but became increasingly negative with each contact (Fig. 8c). Note that
the initial bounce was passive, but from the second bounce the system became active. Only

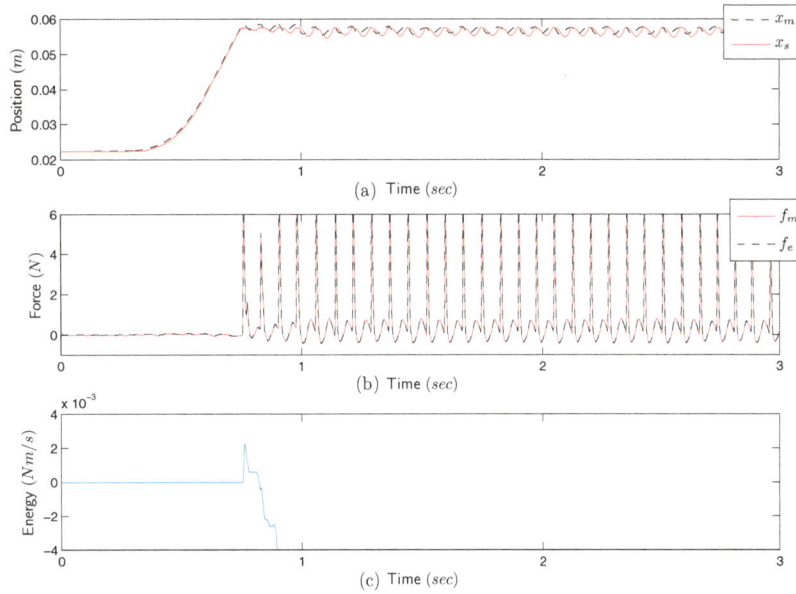

FIGURE 8. Experimental results: hard contact with high velocity (about 120 mm/s). Passivity Controller is inactive (bottom trace) and system exhibits sustained contact oscillations.

X-directional signals are plotted since the main interaction occurs on the X-Axis. The energy in Fig. 8c (and the followed figures with suffix "c") means net supplied energy to the bilateral controller and the PC, and this energy equals to the sum of the value of the PO and the dissipation amount of the PC $E_{obsv}(n) + \alpha_1(n)v_m^2(n) + \alpha_2(n)v_s^2(n)$.

In the next experiment, with the PC turned on, the operator approached the contact point at the same velocity (Fig. 9a), but stable contact was achieved with about 7 bounces (Fig. 9b). Again the first bounce behaved passively, but subsequent smaller bounces were active (Fig. 9c). On the second bounce, the PC at the master port began to operate (Fig. 9d), and eliminated the oscillation by modifying the master control force (Fig. 9b). On the other hand, the PC at the slave port only operated on the second bounce (Fig. 9d). Note that, even though the PC changes the master control force, the human can feel smoother force since the PC only operate very briefly at the end of contact, the PC changes the force with exactly that amount necessary to guarantee stability, and the force is filtered and transmitted to the human operator through the master mechanism.

6.2. Behavior during Low Velocity. In this subsection, we studied the behavior of the PC during low velocity. In this experiment, without the PC, the operator maneuvered the master to make the slave contact the hard wall at about 30 mm/s, and followed a slanted wall in the Y-direction. Since the operator approached contact with relatively low velocity, stable contact and surface following is achieved before $t = 1.5 \ sec$. However, after $t = 1.5 \ sec$, the contact became unstable, resulting in an increasing oscillation (Fig. 10a, b); the value of the PO remained positive before $t = 1.5 \ sec$, but became increasingly negative with each contact.

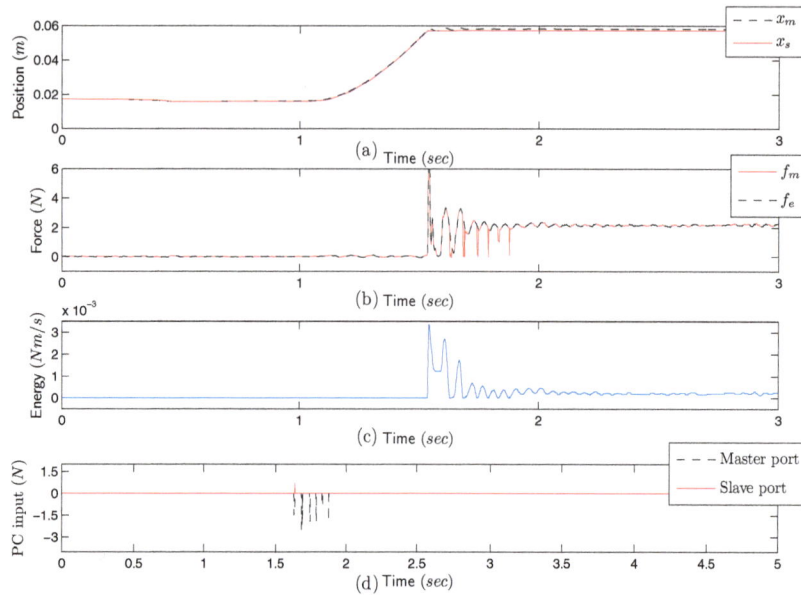

FIGURE 9. Experimental results: hard contact with the same velocity as in Fig. 8 with Series Passivity Controller operating. Oscillation is suppressed by brief pulses of force from Passivity Controller (bottom trace).

In the next experiment, with the PC turned on, the operator maneuvered the teleoperator in the same way, but a stable surface following (Fig. 11a, b) was achieved. The small force bounces before $t = 1.7\ sec$ can be seen to behave passively, but the following small force bounces were active. After $t = 1.7\ sec$, the PC at the master and slave port began to operate (Fig. 11d), and eliminated the oscillation by modifying the transmitted force to the operator (Fig. 11b).

We also tested the system in contact with soft, sponge material. As expected, contact was stable in both cases. However, the value of the PO crossed to negative values for some very brief intervals (100-200 $msec$). Note that, although many applications of teleoperation involve hard contact, surgery application often involves contact with soft materials.

7. Conclusions and Future Works

In this Chapter, a general method for implementing the time-domain passivity control scheme to teleoperation systems is introduced. Any existing teleoperation system, having arbitrary bilateral control architecture, can be stabilized with several additional lines of controller code. Design options exist for how to distribute energy dissipation for the case where energy is produced by two ports.

Since, to maintain stability, the PC only degrades performance (through the added damping of the PC) when it is needed, and only in the small amount needed, the performance can be maximized based on the guaranteed stability. Note that the proposed

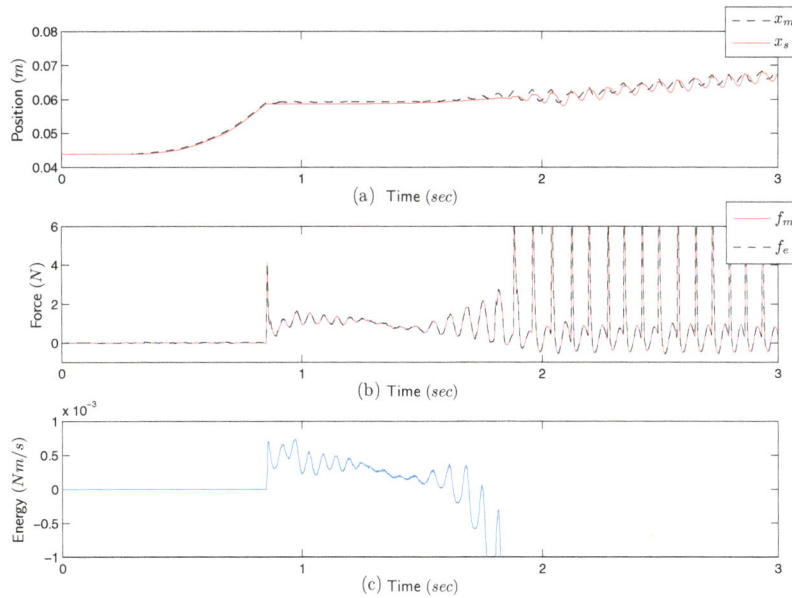

FIGURE 10. Experimental results: hard surface following after slow contact. Passivity Controller is inactive (bottom trace) and system exhibits sustained oscillations during the following.

controller is not a method to increase performance (transparency), but a method to preserve performance while guaranteeing stability of a high performance system by adding the PO/PC to a conventional bilateral control scheme. The PC is expected to be very useful in teleoperation since a known model or parameter estimation are not required.

There are several areas of future work that need to be pursued. The first issue is the identification of an external dissipation amount and subsequently using it for the design of the PO/PC. In teleoperation systems, we need to real-time estimate the dissipation amount of the human/environment and master/slave mechanism that may allow the PC to operate with a different threshold and thus give even less conservatism.

In addition to the threshold problems, a method to apply the PO/PC to teleoperation systems with communication time-delay must also be resolved. Due to the time-delay, it is difficult to monitor the energy flow in and out of the bilateral controller in real-time software.

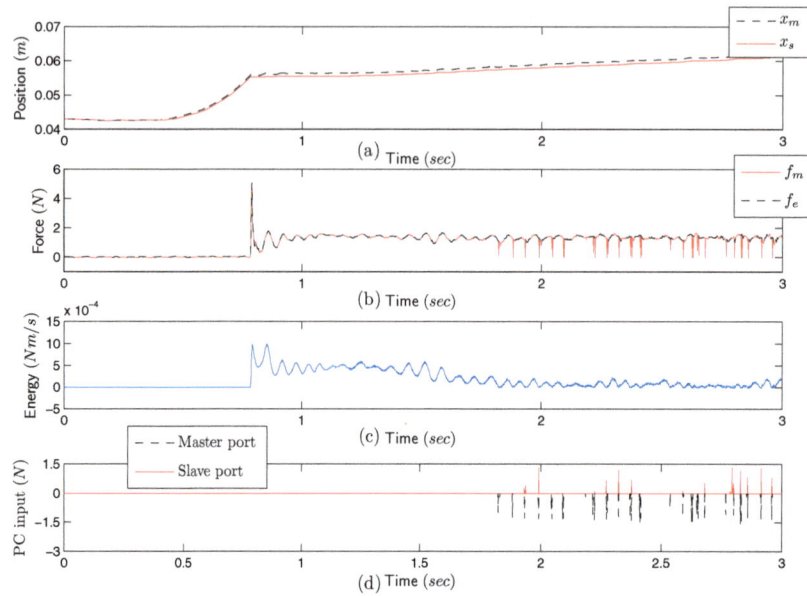

FIGURE 11. Experimental results: hard contact with the same velocity as in Fig. 10 with Series Passivity Controller operating. Oscillation is suppressed by brief pulses of force from Passivity Controller (bottom trace).

Bibliography

[1] R. J. Adams and B. Hannaford, "Stable Haptic Interaction with Virtual Environments," *IEEE Trans. Robotics and Automation*, vol. 15, no. 3, pp. 465-474, 1999.

[2] R. J. Anderson, "Autonomous, Teleoperated, and Shared Control of Robot Systems," *Proc. IEEE Int. Conf. Robotics and Automation*, Minneapolis, MN, pp. 2025-2032, 1996.

[3] R. J. Anderson and M. W. Spong, "Asymptotic Stability for Force Reflecting Teleoperators with Time Delay," *Int. Journal of Robotics Research*, vol. 11, no. 2, pp. 135-149, 1992.

[4] J. E. Colgate, "Robust Impedance Shaping Telemanipulation," *IEEE Trans. Robotics and Automation*, vol. 9, no. 4, pp. 374-384, 1993.

[5] J. E. Colgate and G. Schenkel, "Passivity of a Class of Sampled-Data Systems: Application to Haptic Interfaces," *American Control Conference*, Baltimore, MD, pp. 3236-3240, 1994.

[6] C. A. Desoer and M. Vidyasagar, *Feedback Systems: Input-Output Properties*, New York: Academic, 1975.

[7] B. Hannaford, "Design Framework for Teleoperators with Kinesthetic Feedback," *IEEE Trans. Robotics and Automation*, vol. 5, no. 4, pp. 426-434, 1989.

[8] B. Hannaford and J. H. Ryu, "Time Domain Passivity Control of Haptic Interfaces," *IEEE Trans. Robotics and Automation*, vol. 18, no. 1, pp. 1-10, 2002.

[9] N. Hogan, "Controlling Impedance at the Man/Machine," *Proc. IEEE Int. Conf. Robotics and Automation*, Scottsdale, AZ, pp. 1626-1631, 1989.

[10] K. Hashtrudi-Zaad and S. E. Salcudean, "Adaptive Transparent Impedance Reflecting Teleoperation," *Proc. IEEE Int. Conf. Robotics and Automation*, pp. 1369-1374, 1996.

[11] K. Hashtrudi-Zaad and S. E. Salcudean, "On the use of Local Force Feedback for Teleoperation," *Proc. IEEE Int. Conf. Robotics and Automation*, pp. 1863-1869, 1999.

[12] K. Hashtrudi-Zaad and S. E. Salcudean, "Analysis of Control Architectures for Teleoperation Systems with Impedance/Admittance Master and Slave Manipulators," *Int. Journal of Robotics Research*, vol. 20, no. 6, pp. 419-445, 2001.

[13] H. Kazerooni and C. L. Moore, "An Approach to Telerobotic Manipulations," *ASME Journal of Dynamic Systems, Measurement, and Control*, vol. 119, pp. 431-438, 1997.

[14] D. A. Lawrence, "Stability and Transparency in Bilateral Teleoperation," *IEEE Trans. Robotics and Automation*, vol. 9, no. 5, pp. 624-637, 1993.

[15] D. Lee and P. Li, "Passive Coordination Control for Nonlinear Bilateral Teleoperated Manipulators," *Proc. IEEE Int. Conf. Robotics and Automation*, pp. 3278-3283, 2002.

[16] D. Lee and P. Li, "Passive Feedforward Approach to the Control of Bilateral Teleoperated Manipulators," *ASME Haptics Symposium at IMECE*, 2000.

[17] H. Lee and M. J. Chung, "Adaptive Controller of a Master-Slave System for Transparent Teleoperation," *Journal of Robotic Systems*, vol. 15, no. 8, pp. 465-475, 1998.

[18] G. M. H. Leung, B. A. Francis, and J. Apkarian, "Bilateral Controller for Teleoperators with Time delay via μ-Synthesis," *IEEE Trans. Robotics and Automation*, vol. 11, no. 1, pp. 105-116, 1997.

[19] R. Lozano, N. Chopra, and M. W. Spong, "Passivation of Force Reflecting Bilateral Teleoperators with Time Varying Delay," *Mechatronics'02*, Entschede, Netherlands, pp. 24-26, 2002.

[20] G. Niemeyer and J. J. Slotine, "Stable Adaptive Teleoperation," *IEEE Journal of Oceanic Engineering*, vol. 16, pp. 152-162, 1991.

[21] J. H. Ryu and D. S. Kwon, "A Novel Adaptive Bilateral Control Scheme using Similar Closed-loop Dynamic Characteristics of Master/Slave Manipulators," *Journal of Robotic Systems*, vol. 18, no. 9, pp. 533-543, 2001.

[22] S. E. Salcudean, "Control for Teleoperation and Haptic Interfaces," *Control Problems in Robotics and Automation*, B. Siciliano and K.P. Valavanis (Eds), Springer-Verlag Lecture Notes in Control and Information Sciences, vol. 230, pp. 50-66, 1997.

[23] S. E. Salcudean, K. Hashtrudi-Zaad, S. Tafazoli, S. P. DiMaio, and C. Reboulet, "Bilateral Matched-Impedance Teleoperation with Application to Excavator Control," *IEEE Control Systems Magazine*, vol. 19, no. 6, pp. 29-37, 1999.

[24] A. J. Van der Schaft, "L2-Gain and Passivity Techniques in Nonlinear Control," *Springer, Communications and Control Engineering Series*, 2000.

[25] J. C. Willems, "Dissipative Dynamical Systems, Part I: General Theory," *Archive for Rational Mechanics and Analysis*, 45, pp. 321-351, 1972.

[26] J. Yan and S. E. Salcudean, "Teleoperation Controller Design Using H_∞-Optimization with Application to Motion-scaling," *IEEE Trans. Control Systems Technology*, vol. 4, no. 3, pp. 244-258, 1996.

[27] Y. Yokokohji and T. Yoshikawa, "Bilateral Control of Master-slave Manipulators for Ideal Kinesthetic Coupling-Formulation and Experiment," *IEEE Trans. Robotics and Automation*, vol. 10, no. 5, pp. 605-620, 1994.

[28] Y. Yokokohji, T. Imaida and T. Yoshikawa, "Bilateral Control with Energy Balance Monitoring Under Time-Varying Communication Delay," *Proc. IEEE Int. Conf. Robotics and Automation*, San Francisco, CA, pp. 2684-2689, 2000.

[29] W. Zhu and S. E. Salcudean, "Stability Guaranteed Teleoperation: An Adaptive Motion/Force Control Approach," *IEEE Trans. Automatic Control*, vol. 45, no. 11, pp. 1951-1969, 2000.

INDEX